D0906991

OPTOELECTRONICS

OPTOELECTRONICS

Robert G. Seippel

RESTON PUBLISHING COMPANY, INC.
A Prentice-Hall Company
Reston, Virginia

Library of Congress Cataloging in Publication Data

Seippel, Robert G
   Optoelectronics.

   Bibliography:
   Includes index.
   1. Optoelectronics.   I. Title.
TA1750.S44   621.36   80-22288
ISBN 0-8359-5255-X

© 1981 by Reston Publishing Company, Inc.
A Prentice-Hall Company
Reston, Virginia 22090

3 5 7 9 10 8 6 4 2

Printed in the United States of America

TABLE OF CONTENTS

PREFACE

Optoelectronics has suddenly become the dominant growth area in the electronics field. From light emitting and receiving devices through fiber optics to lasers, optoelectronics must now be dealt with as an emerging giant of the industry.

Devices such as light emitting diodes (LEDs), injection laser diodes (ILDs), optoisolators, and detectors are being built stronger, in more and different sizes, shapes, and colors and with more current carrying capacity. Displays are now manufactured with built-in intelligence and computer capabilities.

Fiber optics is no longer an infant in the electronics field. It has grown from a laboratory science into a practical tool for transmission of information. Fiber optic electronics is spreading rapidly to aerospace, heavy equpment, medicine, mining, oceanography, and most other technical fields.

Just what are fiber optics and why this rapid expansion? This new technology converts electrical impulses into light energy and transmits the light waves over an extremely small fiber to a receiver. At the receiver, the light energy is converted back to electrical impulses to perform its original task.

Why use fibers instead of electrical wire? First of all, because fibers are extremely light and small. Secondly, they are virtually immune to electrical interference. The fibers can be directed past high voltage areas. They will neither radiate outward the signal they carry, nor are they susceptible to the acceptance of induced signals. Finally, optical fibers will accept a large bandwidth with extremely low power loss. Altogether, efficient advanced technology.

Much of the credit for the phenomenal growth of fiber optics is due to the pioneers in the development of glass fibers, such as Standard Telecom Labs in England, Nippon Sheet Glass Company in Japan, and Bell Telephone Labs and

Corning Glass Works in the United States. Since 1976, advancements in the fiber optics field have come so quickly that it is difficult to determine who developed what. Dozens of companies in the United States are creating new fibers and connectors, new light launching devices, and related electronics. A momentous new technological breakthrough seems highly unlikely, simply because industry is already there. The future promises improved manufacturing techniques that will tend to bring prices down to meet the needs of consumer economics.

While the world was thinking of the laser as a mysterious Buck Rogers type ray gun, Charles H. Townes of Columbia University was developing the first laser in 1954. Later in 1960, Theodore H. Maiman of Hughes Aircraft constructed the first ruby laser. In the closing months of 1960, A. Javin of Bell Labs created the first gas laser. Semiconductor lasers were devised simultaneously by General Electric, IBM, and Lincoln Laboratories in 1962.

These pioneers set the pace and companies such as Coherent, Molectron and Spectra-Physics have followed by raising laser technology to star status in industry.

Today, lasers are no longer fictional aspirations. The broadest possible applications for lasers are now being realized. Some of these applications are in communications holography, medicine, direction finding, manufacturing, and light transmission.

Optoelectronics is old and new. It still has mystique but that mysteriousness is now mixed with reality. This book is dedicated to pioneers in the field and to the many who are dedicating their efforts toward its future.

*   *   *

A special thanks is given to the contributors of data and illustrations within this book. I have attempted to include everyone who has assisted me in this publication endeavor. Those whose articles I have used as reference are listed in the bibliography. Illustrations have the standard courtesy lines. I sincerely hope that I have left no one out, for without the contributors' help this book would not be possible.

Robert G. Seippel, Ph.D.

AMP Inc., Harrisburg, Pennsylvania
   Walter Scott, Marketing
   Dennis Bukowski, Product Manager
AT&T, New York, New York
   Dave Manahan, Public Relations
Canoga Data Systems, Canoga Park, California
   Richard C. McCaskill, Product Manager
Coherent, Palo Alto, California
   Edwina DeRousse, Manager, Marketing Communications

Corning Glass Works, Corning, New York
  Arthur N. Martin, Public Relations
Department of the Air Force, Headquarters
    Electronics Systems Division (AFSC), Hanscom Air Force Base,
      Massachusetts
  Mary Farris, Public Information
Department of the Air Force, Rome Development Center,
  Griffis Air Force Base, New York
  Juventino R. Garcia, Editor
General Cable, Greenwich, Connecticut
  John Cavanaugh, Advertising and Sales Promotion
General Telephone and Electronics Corp., Santa Monica, California
  Chuck Halloran, Public Affairs
  John Lawrence, Public Affairs
ITT Cannon Electric, Santa Ana, California
  Dick Harmon, Public Relations
ITT Electro-Optical Products Div., Roanoke, Virginia
  Bob Blanchard, Public Relations
ITT Telecommunications, Raleigh, North Carolina
  Patricia Mullen Grant, Manager Public Relations and Advertising
Litronix Inc., Cupertino, California
  Doug Fraser, Special Accounts Manager
Molectron Corp., Sunnyvale, California
  Jean E. Pellegrino, Manager, Sales Operations
Phase-R Corp., New Durham, New Hampshire
  Paul M. Prause, Marketing
New York Telephone, New York, New York
  Mike Clendenin, Public Affairs
Siecor Optical Cables Inc., Horseheads, New York
  Marcia Van Etten, Marketing Communications
Spectra-Physics, Mountain View, California
  Bob Yazell, Marketing
United Detector Technology Inc., Santa Monica, California
  Ira Wagner, Sales Manager, Detectors
Western Electric, New York, New York
  Saul Fingerman, Technical Information

ABBREVIATIONS USED IN OPTOELECTRONICS

Abbreviation	Identity
AGC	Automatic Gain Control
AMI	Alternate Mark Inversion
APD	Avalanche Photodiode
AVC	Automatic Volume Control
BeO	Beryllium Oxide
BER	Bit Error Rate
CGS	Centimeter-Gram-Second System
cm	centimeter
CO_2	Carbon Dioxide
CVD	Chemical-Vapor Deposition
CW	Continuous Wave
E	Energy
E-M	Electromagnetic
EMI	Electromagnetic Interference
EMP	Electromagnetic Pulse
FDM	Frequency Division Multiplexing
FO	Fiber Optics
HCN	Hydrogen Cyanide
He	Helium
HeCd	Helium Cadmium
HeNe	Helium Neon
Hg	Mercury
Hz	Hertz
IR	Infrared
Iv	Luminous Intensity
kW	kilowatts
LASER	Light Amplification by the Stimulated Emission of Radiation

Abbreviation	*Identity*
LED	Light Emitting Diode
MKS	Meter-Kilogram-Second System
mm	Millimeters
Mv	Luminous Exitance
mW	Milliwatts
NA	Numerical Aperture
Nd	Neodyium
Ne	Neon
nm	Nanometers
NRZ	Nonreturn to Zero
PCM	Pulse Code Modulation
PCS	Plastic Clad Silica
PF	Packing Fraction
PIN	Positive Intrinsic Negative Layers
PPM	Pulse Position Modulation
PRF	Pulse Rate Frequency
PRM	Pulse Rate Modulation
PRR	Pulse Repetition Rate
PRT	Pulse Repetition Time
PWM	Pulse Width Modulation
Q	Quality of a Laser Beam
Qv	Luminous Energy
RFI	Radio Frequency Interference
RMS	Root Mean Square
RZ	Return to Zero
SNR	Signal-to-Noise Ratio
TDM	Time-Division Multiplexing
TEM	Transverse Electromagnetic Mode
UV	Ultraviolet
YAG	Yttrium Aluminum Garnet

1

INTRODUCTION
TO OPTOELECTRONICS

PROGRESSION OF THE THEORIES OF LIGHT

The earliest theory of light was a theory of vision. It was thought that the eyes sent out invisible antennae and were thus able to feel or sense those things that were too distant to touch. This was called the "tactile" touching theory.

A second theory was the "emission" theory. This viewed all objects as emitting something which, when entering the eye, was sensed by some part of the eye, therefore the brain could see. Both the tactile theory and the emission theory were accepted by the Greek thinkers in 500 B.C. The tactile theory was dismissed, but the emission theory carried on and was used by Newton in developing his corpuscle theory.

Robert Hooke, a noted British scientist, published a book called *Micrographia* in 1665. In it he described light as small rapid vibrations and that variations in color are produced by these vibrations, thus originating the wave theory of light. The Royal Society of Great Britain honored Hooke by making him curator of experiments. Then in 1672, Sir Isaac Newton sent the society an account of his experiments in color separation with the prism. Shortly after the color theory, Newton developed his corpuscle theory and sent an account of it to the Royal Society. It was immediately turned down by the committee, led by Robert Hooke, because it conflicted with Hooke's wave theory.

As early as 1676, a Danish astronomer, Olaus Roemer, discovered that the speed of light was finite (measureable) and estimated its speed at 140,000 miles per second. It was not until 1926 at Mt. Wilson that Albert Michelson, an American, determined through more thorough experiments that the speed of light is 299796 $\pm$ 4 kilometers per second. The speed of light has been estimated by so many scientists that it would not be meaningful to present them all.

Along with the fact that light has velocity, experiments by Christian Huygens, a seventeenth-century Dutch scientist, proved that light also had polarity (or direction) depending on the plane. However, it wasn't until the nineteenth century that detailed experiments were made with light reflection and refraction. The two findings (polarity and finite speed of light), along with several wave theories, further led to the theory of transverse waves. It was thought that an elastic medium pervaded all space and only changed when matter was in its way.

In other words, light was a thing that moved at a finite speed, was polarized, and could be modified by refraction or reflection. In his famous principle in 1690 Christian Huygens stated that each point of a wavefront may be regarded as a source of new waves. This was very much like Maxwell's later electromagnetic theory of light. James Clerk Maxwell, a Scot, through experiments with propagation of electromagnetic waves in the middle 1800's, derived mathematical constants for electricity and magnetism. By these constants he was able to establish a frequency and wavelength electromagnetic spectrum. The Maxwell mathematical constants of the electromagnetic wave theory allowed Heinrich Hertz, a German physicist, to produce wavelengths by electrical means in 1887.

In 1900, Max Planck found that light energy is emitted by atoms in multiples of an energy unit. The unit was called a quantum and its magnitude depended on wavelength. Seven years later, in 1907, Einstein found concentrations of energy, which he called photons, when light released electrons from atoms. They were propagated like particles from the electrons.

Modern ideas of relativity, at least in the modern age are due to the work of Albert Einstein. The modern theory of gravitation is general relativity and was published by Einstein in 1915. General relativity pictures the world in four dimensional space/time. Particles and light rays move along paths in four dimensional space in the straightest possible curves. The curves are dependent on the matter present in the space. All reference to absolute three dimensional space are absent. General relativity is still largely theoretical. Special relativity asserts that all laws of physics are equally valid in all inertial frames of reference. In 1905 Einstein published this special theory of relativity. The importance of this publication is that the theory has some extremely striking practical implications. The first implication is that time and space separations are dependent on the selection of motion of the observer. The second prediction is that no body or physical effect travels faster than the speed of light. The third and probably the most known is the relation implication of mass m, energy E, and the speed of light c. The impact of the formula $E = mc^2$ is phenomenal. Technology has made great strides in areas such as atomic power and partical accelerators. Special relativity is by far the most practical and valid of all branches of physics.

Modern quantum mechanical light theory began around 1927 and incorporates the appropriate parts of the electromagnetic wave theory, the quantum theory, and the relativity theory.

In figure 1–1, the progression of major light theories is presented in chronological order. Minor theories have been omitted to preserve simplicity.

Figure 1–1 Progression of the Theories of Light (*Mike Anaya*)

PROGRESSION OF PHOTOELECTRICITY

The progression of photoelectricity necessarily follows the progression of light. The probable beginning of photoelectricity was with the discovery of the photovoltaic cell by Alexandre Becqueral in 1839. Becqueral found that a battery he was work-ing with developed a better charge when exposed to light.

In 1888, Heinrich Hertz working on the invisible portion of the spectrum revealed the fact that certain metals produced electrical effects when exposed to ultraviolet light. The next steppingstone to photoelectricity was Willoughby Smith's discovery that the resistance of selenium changes when light is applied. Among other scientists experimenting with this phenomena in the late nineteenth century was R. E. Day.

Of extreme importance in photoelectricity history and indeed science in general was the discovery of radiation in 1896 by Antoine Henri Becqueral, son of Alexandre Becqueral the discoverer of the photovoltaic cell. The Becqueral family's work with light and electricity must be lauded as some of the greatest achievements of the time.

Just after the turn of the century, about 1905–1906, Max Planck de-veloped his universal mathematical constant that light is emitted by atoms in multiples of a certain unit or quanta. The size of the unit being a quantum, which depends on the wavelength of the radiation. Einstein, in 1907, suggested that in order to give an adequate description of these localized pockets of energy de-veloped from light exposure, that they had to be considered as particles separate from the atoms. Thus the concept of particles or photons of light energy was developed. In 1927, the photon theory was added to the quantum theory, the theory of relativity, and the electromagnetic theory, to establish the present-day light theory called quantum mechanics previously discussed. Figure 1–2, shows the development of the photoelectrical theory.

MODERN PROGRESS IN ELECTRONICS

Throughout the 1930s and 1940s many laboratories such as Bell Telephone and General Electric were developing photoelectric devices for commercial and mili-tary uses. The major breakthrough for the world of electronics was the invention of the transistor by Bardeen, Brattain, and Shockley of Bell Labs in 1948. This was the beginning of the age of electronics called *solid-state* or *semiconductor electronics*.

Since this discovery, electronics has been widely advanced by the design of integrated circuits. The principal designer was Kilby of Texas Instruments in 1958. Since integrated circuits are such a vast field in electronics today, we could not possibly give credit to everyone who is responsible for advancement in this area.

A stage in optoelectronics which seems appropriate to mention is the

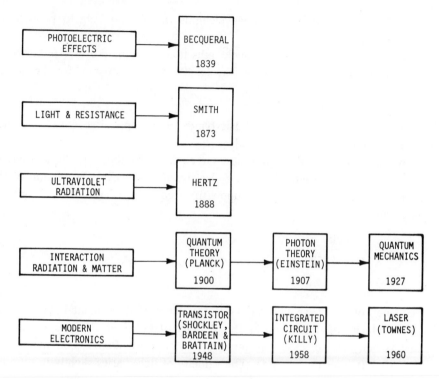

Figure 1–2 Progression of Photoelectricity (*Mike Anaya*)

ultimate use of optoelectronics, the laser or "Light Amplification by Stimulated Emission of Radiation." In 1958, Charles H. Townes, an American, and Arthur L. Schawlow, a Canadian, wrote a paper in which they showed that it was possible to use stimulated emission for amplifying light waves as well as microwaves (the maser). Two years later, in 1960, Theodore H. Maiman, a scientist working at Hughes Aircraft Company in California built the first optical maser, otherwise known as the laser.

OPTOELECTRONIC OPERATIONAL SPECTRUM

Optoelectronics is the branch of electronics that deals with light. Electronic devices involved with light operate within the optical part of the electronic-frequency spectrum. The visible range is only a narrow portion of the optical spectrum. There are three basic bands of the optical frequency spectrum. These are:

1. Infrared—Band of light wavelengths that is too long to be seen by the human eye.
2. Visible—Band of light wavelengths that the human eye responds to.
3. Ultraviolet—Band of light wavelengths that is too short for the human eye to see.

Before we discuss each of the three basic bands of the optical spectrum, we shall explain several things that are involved in it.

A wavelength λ (the lower-case Greek letter lambda) is the amount of space occupied by the progression of an electromagnetic wave. Calculation of the size of the wavelength is dependent upon the frequency of the wave f and the velocity of light c.

$$\lambda = \frac{c \text{ (velocity of light)}}{f \text{ (frequency)}}$$

$$\lambda = \frac{c}{f}$$

The frequency of the wave f is the same frequency that originated the radiated wave. The velocity of the wave c is the velocity of light, which is approximately the same velocity for all electromagnetic waves, that is 300,000,000 meters per second (186,000 miles per second). The velocity of light varies slightly in different materials.

An audio frequency of 15,000 hertz (cycles per second) would have a wavelength of

$$\lambda = \frac{300,000,000}{15,000} = 20,000 \text{ meters}$$

A radio frequency of 300 megahertz would have a wavelength of

$$\lambda = \frac{300,000,000}{300,000,000} = 1 \text{ meter}$$

Larger frequencies such as 3,000 megahertz would have a wavelength of 0.1 meter.

Even larger frequencies such as 300,000 megahertz would radiate a wavelength of 0.001 meter, which could also be written as 1000 microns. It is convenient to discuss smaller wavelengths in microns because of the large size of the frequency involved.

In the optical spectrum another value Å (the angstrom) is used to express wavelength.

1 angstrom (Å) = 0.0001 micron (μ)

or

$$1 \text{ micron } (\mu) = 10,000 \text{ angstrom } (\text{Å})$$

In terms of scientific notation

$$1 \text{ angstrom } (\text{Å}) = 10^{-10} \text{ meter}$$
$$1 \text{ micron } (\mu) = 10^{-6} \text{ meter}$$

The optical spectrum operates from wavelengths of 0.005 microns to 4000 microns (50 angstroms to 40,000,000 angstroms). In frequencies, these are extremely high values (6×10^{16} hertz to 7.5×10^{10} hertz). (See figure 1–3.)

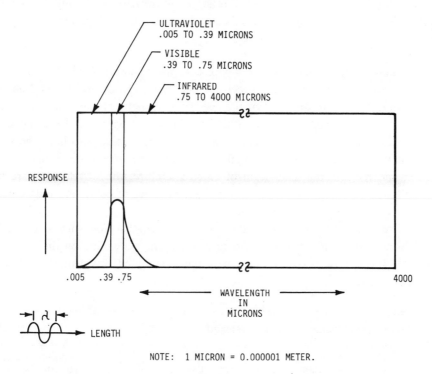

Figure 1–3 The Optical Spectrum Including the Portion Visible to the Human Eye (*Mike Anaya*)

Ultraviolet Wavelengths Wavelengths for ultraviolet radiated frequencies range from 0.005 microns to 0.3900 microns (50 to 3900 angstroms). In frequencies, these are 6×10^{16} hertz to 7.69×10^{14} hertz.

Visible Wavelengths Visible radiated wavelengths range from 0.3900 microns to 0.7500 microns (3900 to 7500 angstroms). In frequencies, these are 7.69×10^{14} hertz to 4×10^{14} hertz.

Infrared Wavelengths   Wavelengths for infrared radiated frequencies range from 0.7500 microns to 4000 microns (7500 to 40,000,000 angstroms). In frequencies, these are 4×10^{14} hertz to 7.5×10^{10} hertz. Infrared frequencies are too low and ultraviolet frequencies are too high for human eye response.

Wavelengths are in indirect proportion to frequency, therefore infrared wavelengths are too long for human-eye response and ultraviolet wavelengths are too short for human-eye response. Another point worth mentioning is that the wavelength range for infrared (0.75 to 4000 microns) is much greater than either visual or ultraviolet wavelength ranges.

Color   The human eye sees violet (approximately 0.43-micron wavelength) on one side of the color spectrum and red (approximately 0.68-micron wavelength) on the opposite side of the spectrum. In between these extremes, the eye sees blue, green, yellow, and orange. Two beams of light that have the same wavelength are seen as the same color. Two beams that are seen to have the same color usually have the same wavelength. However, a mixture of two colors such as green and red will match a beam of blue. Therefore, the wavelengths are not necessarily the colors they may seem to the eye. Matches of colors are only correct to the person who has extremely good vision.

The graduation of colors and their wavelengths from one end of the color (visual) spectrum have overlapping wavelengths depending on the observer. Table 1–1 shows typical colors along with their estimated wavelengths and radiated frequencies.

TABLE 1–1   Optical Colors and Their Wavelengths

	COLOR	WAVELENGTHS (IN MICRONS)*	FREQUENCIES (IN HERTZ)
	Ultraviolet	0.005–0.39	6×10^{16}–7.69×10^{14}
	Violet	0.40 –0.45	7.5 –6.6×10^{14}
Visual	Blue	0.45 –0.50	6.6 –6.0×10^{14}
Response	Green	0.50 –0.57	6.0 –5.27×10^{14}
	Yellow	0.57 –0.59	5.27–5.01×10^{14}
	Orange	0.59 –0.61	5.01–4.92×10^{14}
	Red	0.61 –0.70	4.92–4.28×10^{14}
	Infrared	0.70 –4000	4.28×10^{14}–7.5×10^{10}

* Approximates only, with overlapping wavelengths and frequencies.

LIGHT RAYS

Light travels in beams or rays. A ray is the path of the light. In the seventeenth century, propagation of light was simply explained and represented by rays. In figure 1–4, light from a pinpoint-size hole is interrupted by an opaque (not pervious to light) disc in its path. The opaque disc prevents light from passing through. This causes a distinct, sharp-edged shadow on a backboard receptor or movie screen. The small hole shadow is called an *umbra*. This figure is an extremely good example to argue the facts that light travels from a source to a receptor on beams or rays and secondly that the rays are straight lines. Rays interrupted by opaque objects cannot get to the receptor. Newton's first laws of motion stated that particles travel in a straight line unless moved by some force. Newton therefore determined that light particles were weightless and furthermore the particle's speeds were so fast that gravity had no effect on them. This is easy to accept as illustrated in the figure 1–4.

Figure 1–4 Formation of Shadows in Support of Straight-Line Propagation of Light (*Mike Anaya*)

Propagation Sources Light rays on beams are propagated at various wavelengths. Some sources radiate a much narrower bandwidth than others. (See figure 1–5.) Monochromatic light is radiation that has one wavelength (or an extremely narrow band of several wavelengths). Heterochromatic light is radiation that has several very distinct wavelengths. Panchromatic light is radiation that includes wavelengths in a great range of the spectrum.

Action of Light on Material All material does not "see" light in the same manner. There are basically three different light-receptor materials. These are

Figure 1–5 Radiation from Sources of Light Rays (*Mike Anaya*)

opaque, transparent, and translucent. Material that does not allow any light to travel through it is called *opaque*. If you can see light partially through a substance but not clearly, the substance is called *translucent*. Finally, *transparent* material allows you to see through it clearly.

Photons   Response of the human eye to light is determined by the energy of the radiation. Radiant energy travels in energy groups known as *photons*. The frequency of the radiation determines the strength of the photons. The higher the frequency of the radiation, the higher the energy of the photons. As an example, wavelengths in the visible range of the spectrum near the ultraviolet range have short wavelengths, higher frequencies, and higher photon energy. Furthermore, wavelengths in the visible range of the spectrum near the infrared range have long wavelengths, lower frequencies, and lower photon energy. When an electron is freed by absorbing the energy of a photon the result of the combination is called a photoelectron. Again the energy of the photoelectron is dependent on the frequency of the radiation. The energy of a photon is expressed in watts and can be calculated by the following equation:

$$W = hf$$

where W = energy in watts

h = Planck's constant (6.63×10^{-34} joule-seconds)

f = frequency of the photon wavelength in hertz (Hz)

Diffraction   Up to this point we have assumed that light travels in straight lines. This is not altogether true. In experiments by Newton called *Newton's rings,* he discovered that light does spread, though to a very small extent, from the edges of the light beam. The shadow of an opaque obstacle is not perfectly sharp, but some light does indeed penetrate into the shadow area. The fringes near the opaque shadow shown in figure 1–4 appear sharp, but under magnification would show interference or what we now call *diffractions*. In conclusion, we might add that light does not always correspond to its visual appearance.

LENS SYSTEM

A lens is a piece of glass or other transparent device such as plastic, with two curved or one curved and one flat surface. The purpose of the lens is to bring together or spread apart light rays as they pass through it. There are two basic lens shapes, *convex* and *concave*. All lenses are made in the form of one of these shapes or both in combination. We shall illustrate the more simple forms and describe their operation. Lens discussion is purposely brief as the intent is to provide basic operation only. Lenses and their operation could easily fill a text of their own.

Convex Lens   The convex lens is a converging lens. That is, it causes light to come together. In figure 1–6A, a plano convex lens accepts parallel light rays and converges them into a focal point. In figure 1–6B, a plano convex lens accepts light rays from a converged source then collimates the rays or makes them parallel. In figures 1–6C, and 1–6D, a double convex lens does essentially the same thing.

Concave Lens   The concave lens is a diverging lens. That is, it causes light to spread apart. In figure 1–7A, a plano concave lens causes parallel light rays to diverge or spread out. In figure 1–7B, a plano concave lens causes rays that are spread out to become parallel. In figures 1–7C, and 1–7D, a double concave lens does essentially the same thing.

Lenses in Combination   Often it is necessary to utilize two lenses to provide a final result. In figure 1–8, a concave lens is used to convert divergent light rays to parallel light rays. The parallel lines are then directed through a convex lens which converges them into a focal point.

LIGHT SEPARATION

In 1672, Isaac Newton developed his theory of color. He allowed a thin beam of white light to pass through a shutter. He directed this thin beam through a prism. The prism produced colors of the rainbow, red, orange, yellow, green, blue,

(A)   PLANO CONVEX LENS
      CONVERGING LIGHT
      RAYS.

(C)   DOUBLE CONVEX LENS
      CONVERGING LIGHT
      RAYS.

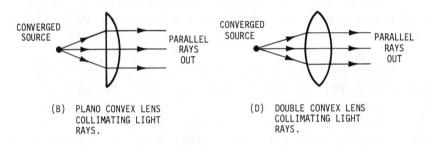

(B)   PLANO CONVEX LENS
      COLLIMATING LIGHT
      RAYS.

(D)   DOUBLE CONVEX LENS
      COLLIMATING LIGHT
      RAYS.

Figure 1–6   Convex Lenses (*Mike Anaya*)

(A)   PLANO CONCAVE LENS
      DIVERGING PARALLEL
      LIGHT RAYS.

(C)   DOUBLE CONCAVE LENS
      DIVERGING PARALLEL
      LIGHT RAYS.

Figure 1–7   Concave Lenses (*Mike Anaya*)

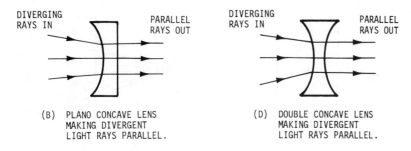

(B) PLANO CONCAVE LENS
MAKING DIVERGENT
LIGHT RAYS PARALLEL.

(D) DOUBLE CONCAVE LENS
MAKING DIVERGENT
LIGHT RAYS PARALLEL.

Figure 1–7 *continued*

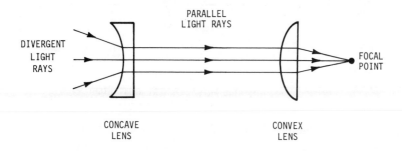

Figure 1–8 Lenses in Combination (*Mike Anaya*)

indigo, and violet as the light passed through. From this he determined that white light is made up of the colors of the spectrum. He further decided that red was bent the least, orange a little more, and so on, to the violet, which was bent the most. Thus he concluded that the angle of refraction through a prism varies with the color.

In figure 1–9A, white light passes through a shutter and a prism to separate the colors of the spectrum. Newton's second thought was to reconstruct the colors back to white light. He placed a second prism next to the first. The colors were changed back to white light as shown in figure 1–9B. Simple lenses are similar to prisms and can focus different-color light at different points, a serious problem in color photography. The problem can be corrected with lens grouping and special coatings.

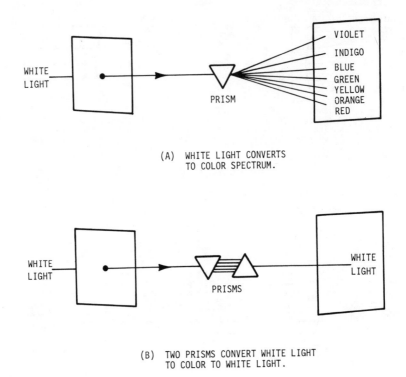

(A)   WHITE LIGHT CONVERTS
      TO COLOR SPECTRUM.

(B)   TWO PRISMS CONVERT WHITE LIGHT
      TO COLOR TO WHITE LIGHT.

Figure 1–9   Light Separation (*Mike Anaya*)

INTRODUCTION TO FIBER OPTIC ELECTRONICS

Fiber Optics Defined   Fiber optics is the science that deals with the transmission of light through extremely thin fibers of glass, plastic, or other transparent material. Optical fibers are dielectric waveguides for electromagnetic energy at optical wavelengths. The fibers provide a path for a single beam of light or in multiples, such as the transposition of a complete image. The fibers are provided as a single fiber or a cable bundle. They may be bent or curved (within limits) to meet the needs of special routing.

History of Fiber Optic Electronics   Probably the founder of fiber-optic electronics was the British physicist John Tyndall. Tyndall found in 1870 that light will follow water into a container and flowing from the container. It wasn't, however, until after World War II and into the 1950s that major corporations began study in the area of light and fiber optics. In 1968, Standard Telecom Labs in England developed silica glass with an extremely low transmission loss. Since then the Corning Glass Works and Bell Telephone Labs in the United States and Nippon Sheet Glass Company in Japan have developed glass fibers with typical loss factors of 1 to 2 decibels per kilometer (dB/km) and even much lower (0.1dB/km) under lab conditions. These low-loss capabilities lend themselves to the development of low-cost, high-efficiency fiber optic communication systems. After 1976, advancements have come so quickly that it is difficult to determine who developed what. Dozens of companies in the United States are creating new fibers and connectors, new light launching devices, and related electronics. A momentous new technological breakthrough seems highly unlikely, simply because industry is already there. The future looks to manufacturing techniques that will bring prices down to meet the needs of consumers.

Optical Fiber Frequency Spectrum   Most optical fibers are manufactured for use in the visible frequency spectrum. Frequency responses tend to fall off on the ultraviolet and the infrared ends of the spectrum. Modern fiber optics have been manufactured to operate in the 4000-angstrom-wavelength range on the ultraviolet range of spectrum. On the infrared end of the spectrum, wavelengths of 10 microns are now common.

Why Use Optical Fibers for Electronic Communication?   The advantages of using fiber optics in communication networks are almost as limitless as a designers ability to create. The optical transmission path provides electrical input and output isolation. To a designer this means total freedom from ground loops, along with lightning-safe installation. Bandwidths are not a problem because they are independent of the cable size. Extremely long-distance cables are available, along with an endless variety of bandwidth products. This means greater data rates at longer distances than wire or coaxial cabling.

One of the major problems encountered in communication electronics is EMI (electromagnetic induction). With fiber optics there is no EMI susceptibility, no induced noise, and no crosstalk. Finally, the fibers and cables are extremely light-weight and low-cost compared to wire. These reasons add up to lower cost in installation and less maintenance time.

Fiber Optics in Operation   In cooperative experiments with Bell Labs in 1973, Western Electric, in its Atlanta, Georgia, facility designed and developed light-guide fibers for telecommunication cabling. The fibers were arrayed into ribbons and placed in experimental cables at Atlantic Bell Labs. These cables were shipped to Chicago for testing. In May of 1978 the AT&T reported a successful one-year trial of a full-service lightwave installation in Chicago. The Chicago test involved a 1.5 mile link under downtown streets between two Illinois Bell Telephone switching stations and between one of the stations and an office housing a number of customers.

In late 1979, AT&T set three more installations into operation. A Connecticut hookup used lightwave technology in a portion of telephone network between the telephone company central office and local terminals located near customer premises. At about the same time a Florida installation provided service to a power station. In Arizona, Mountain Bell supplied service to a government agency near a power station. These experimental stations provided proof that lightwave systems are immune to electrical interference.

Probably New York Telephone set up the most comprehensive and innovative telecommunication systems to date. Two major lightwave installations began operating in 1980. The temporary one in upstate New York was a $2\frac{1}{2}$ mile lightwave link that operated successfully at Lake Placid during the Winter Olympic Games. The second, a permanent installation, links telephone company offices in White Plains with those on East 38th Street in downtown Manhattan. These systems carry incredibly large volumes of information such as voice, video, and computer data and in substantially less space than standard cable and wire systems. Both projects involved Bell Telephone Laboratories, AT&T, Western Electric, and New York Telephone.

In October 1978, General Cable Corporation announced that it would furnish the first optical fiber cable for use with the railroad industry at Union Pacific's Denver, Colorado, terminal. The optical fiber cable was designed and manufactured by General Cable to link a remote CCTV video camera with terminal office equipment to record information, car numbers, and reporting marks printed on passing freight cars. This information is used to check the accuracy of computerized advance makeups of arriving and departing trains, thereby verifying the makeup of the trains as well as the location of all the cars in the trains. The television also provides a check of all cars received from and sent to other railroads in the area. By having accurate information available promptly, switching can be prearranged and speeded up, thereby improving service to shippers. The optical fiber cable is an aerial installation and was continuously monitored for a year in order to assess its performance under conditions of vibration, ice and wind loading, and significant temperature changes. The optical fibers were manufactured by Corning Glass Works, Corning, New York.

Not only must the transmission medium be able to generate reliability over a broad range of harsh environment with a minimum of maintenance but also it must be economically feasible and, ideally, be immune to electrical and electromagnetic interference as well as offer a broad bandwidth of transmission capacity available for other railroad services.

Another optical fiber cable was designed, manufactured, and installed by General Cable for the U.S. Air Force's Arnold Engineering Development Center in Tennessee. The cable connects rocket engine test sites with a central data processing facility for real time data analysis at the free world's largest rocket test facility. This was the first optical fiber cable fully engineered with splices, terminals, and other components for a working high-speed data transmission link—up to 150 megabits per second (Mbps).

For NASA, General Cable designed, manufactured, and installed at Cape Kennedy an optical fiber cable for transmission of high-speed digital data and video signals between locations at the center. They also designed, developed, and manufactured for General Telephone & Electronics Corporation, the world's first optical fiber cable to provide regular telephone service to the public. It was placed in operation in California. Also, a similar optical fiber cable providing telephone service in Brussels, Belgium, was manufactured by General Cable.

Teleprompter Cable Television Incorporated hooked up an 800-foot fiber optic link from an antenna on a Manhattan roof to head-end equipment on the bottom floor of a 34 floor building. The U.S. Navy's Airborne Light Optical Fiber Technology (ALOFT) program successfully demonstrated the use of fiber optics in flight tests on board an A–7 test aircraft. Point-to-point optical links were used. Northrup Corporation and ITT Cannon Electric on a joint research program developed a fiber optic bus to interconnect avionics equipment on board an F–5E tactical fighter. The bus is used on the F–16 and F–18 aircraft.

For several years now a fiber optic telephone system has operated with little or no failure on the *USS Littlerock*. Sonar links are now installed on U.S. navy submarines. The U.S. Army is using optical fiber cables in ground tactical and strategic telecommunications (deployable). General Motors has developed a fiber optic harness system to transmit a control signal to dashboard instruments. Sperry Univac and Digital Equipment Corporation are using fiber optics to connect mainframes and peripheral equipment on their computers.

As early as April, 1976, Rediffusion Engineering Limited in England installed a color TV cable (fiber optic) over a 1.5 kilometer stretch in Hastings, Sussex.

Significant advances have and are being made in fiber optic technology in all electronics fields.

New Innovations In Fiber Optics   Vehicles used to lift personnel into high-tension lines may use fiber optics to reduce the chance of power spikes and the risk of injuring personnel by moving them into high-tension lines.

Information could be taken to miners by fiber optics and therefore reduce the chance of spark generation.

An ocean diver's heartbeat, breath rate, and body temperature could be monitored all on one fiber.

Life-support systems and patient monitoring systems can be developed with clear safety from external interference.

Basic Operation of Fiber Optic Systems A typical fiber optic system is illustrated in figure 1–10. The operation of this simple system is defined in simple terms here and will be described in detail in chapter 8. A system consists of the following:

1. A transmitter accepts an electrical signal and converts it to a current to drive the light source.
2. The light source launches the optical signal into the fiber.
3. The optical fiber provides a path for the optical signal.
4. A light detector detects and converts the optical signal to an electrical signal.
5. A receiver produces low noise and large voltage gain from the power detector signal.
6. Various connectors and splices interface the system.

Figure 1–10   Typical Fiber-Optic System *(Mike Anaya)*

INTRODUCTION TO LASER TECHNOLOGY

Laser Technology Defined Laser technology is the science that deals with the concentration of light into small but powerful beams. The word *laser* is an acronym, that is, it is derived from the initials of *L*ight *A*mplification by *S*timulated *E*mission of *R*adiation. The acronym was chosen when the technology of the maser was shifted from microwaves to lightwaves.

Maser is an acronym for *M*icrowave *A*mplification by *S*timulated *E*mission of *R*adiation. The function of the maser is to concentrate atoms in an active medium such as a crystal or a gas at microwave frequencies, raise them to a higher level of energy by excitation, and then emit this radiated energy into a narrow, intense beam. The function of the laser is the same, except that the operating frequency of the laser is in the optical spectrum, whereas the maser operates in microwave frequencies. The wavelength of microwaves range from 1 millimeter to around 30 centimeters (and probably greater). Optical wavelengths range from 0.005 microns to around 4000 microns. (A micron is equal to 10^{-6} meter.)

History of the Laser   Discussion of maser technology began in the early 1950s, but the first maser was not built until 1954. C. H. Townes, a professor at Columbia University, and students conceived and constructed the first maser. It was developed in an ammonia medium. Solid-state masers using paramagnetic ions imbedded in crystal were developed late in 1957.

The maser researchers were involved in extending the use of masers into the optical wavelength. In 1958 a classic article was published on optical masers by C. H. Townes and A. L. Schawlow, a Canadian. The article triggered enthusiasm which led to the development of the laser. The first laser was created by Theodore H. Maiman, a scientist working for Hughes Aircraft Company in California. Maiman directed a beam of light into ruby crystals with a xenon flash-lamp and measured the emitted radiation from the ruby. He discovered that an increase in input radiation beyond threshold caused emitted radiation to become extremely intense and highly directional. This was ironic, since maser scientists had rejected the ruby as a poor material for masers.

Uranium lasers were developed near the end of 1960 along with other rare-earth materials. Also near the end of 1960, A. Javin of Bell Laboratories created the helium neon laser. Semiconductor lasers (injection laser diodes) were simultaneously manufactured in 1962 by three American companies, General Electric, IBM, and Lincoln Laboratories. Therefore it is difficult to give credit to one.

Laser Types   There are basically four laser types, gas lasers, liquid lasers, solid lasers, and semiconductor lasers. The gas laser uses gas as an active medium. A mixture of helium and neon is enclosed in a glass tube. An electrical current is discharged into the gas and a continuous flow of coherent (one-frequency) waves is issued through the output coupler. The continuous-wave (CW) output is mono-chromatic (one-color).

Liquid lasers use organic dyes enclosed in a glass tube for an active medium. A pump circulates the dye in the tube. A powerful pulse of light excites the organic dye.

Solid lasers use a crystal such as the ruby for an active medium. The ruby crystal is solid and cylindrical. Each end is polished and parallel. A tungsten lamp tied to an alternating current power supply excites the ruby (active medium). The output from the laser is continuous-wave (CW) and coherent.

The semiconductor laser is made from semiconductor pn junctions. It is usually called an injection laser diode (ILD). The excitation mechanism (a direct current power supply) controls the amount of current to the active medium (the ILD). The output light beam is easily modulated.

Laser Characteristics   All lasers have some common characteristics. These are an active material to convert energy into laser light, a pumping source to provide power or energy, optics to direct the beam repeatedly through the active material to become amplified, and optics to direct the beam into a narrow powerful cone of divergence. Other characteristics common to all lasers are a feedback mechanism to provide continuous operation and an output coupler to transmit power out of the laser. These characteristics will be covered in detail in chapter 9.

Laser Light Properties   Laser light has unique properties. The light is one color (monochromatic). Other properties are radiance, directionality, and coherence.

The radiance of the laser is extremely intense and is often compared with the intensity of the sun. The laser, when focused to a fine, hairlike beam, can concentrate all of its power into the narrow beam, therefore it has directional properties. If the beam were allowed to diverge or spread out it would lose its power. Unlike normal beams of light, a coherent beam from a laser will travel long distances with very little divergence.

A laser beam is one-color (monochromatic) and one-frequency, and the light beams are in phase with each other (coherent), that is, they are in time step.

A Basic Laser   As an introduction to laser technology let us take a cursory look at a basic laser, the ruby laser (see figure 1–11). As was previously discussed, the ruby laser was the first developed and was created by Theodore H. Maiman, a scientist at Hughes Aircraft Company. The ruby laser is constructed of an active medium, an excitation source, a feedback mechanism and an output coupler. The active medium in the ruby laser is a tubular crystalline aluminum oxide structure with a small amount of chromium dissolved or doped into its lattice structure. The crystalline structure is grown in the same manner as other solid-state material such as silicon. The ends of the crystal are highly polished and are cut parallel to each other. These ends are coated with a reflective coating. One end of the ruby reflects 100 percent of the light internally. The other end acts as a transmitter, allowing only a small portion of the light to pass outward while reflecting most of the light back into the ruby for stimulated emission. The excita-

Figure 1–11   A Basic Laser (*Mike Anaya*)

tion source is usually a flashtube. The tube is fitted with a trigger that emits a burst of light to start the process.

Let's now go through the operation and consider how it all comes together. The flashtube is energized by a trigger and produces a high-level burst of light similar to a camera electronic flash. The flash, being adjacent to the ruby cylinder, causes the chromium atoms within the crystalline structure to become excited. The process of pumping raises the chromium atoms from ground energy state to excited energy state. The ions then decay and fall to a lower or intermediate energy state. When the population of ions in the intermediate state is greater than the ground state, population inversion is present, that is, there are now more excited atoms than ground-state atoms.

When the population inversion is present, laser action (lasing) can take place. In time, the excited chromium atoms will fall to the ground energy state. When they do, photons are emitted. A photon is a pocket of light caused by radiant energy. The frequency of the energy determines the strength of the photons. Higher frequencies cause greater-strength photons.

Now let us consider the term *stimulated emission*. Chromium atoms are pumped to an excited energy state, then go through a spontaneous decay to an intermediate energy state where an accumulation of excited atoms creates a population inversion. As time progresses, the excited atoms fall from the intermediate energy state to the ground state. When they do, photons are emitted. As the photons are emitted they strike excited atoms and two other identical photons will emerge. These two photons in turn strike other excited atoms and stimulated emission occurs. The more often this happens the greater the intensity of the light beam.

When the beam of photons strikes either mirror ends of the ruby, the reflection of the light beam back into the ruby causes an even greater beam intensity because of photon action. The beam is red in color because the laser is amplifying at a wavelength of 6943 angstroms (0.6943 microns). The white light of the flashtube produces many wavelengths, however only one wavelength can excite the atoms from ground state to excited energy state. The amplified light is fed partially out of the laser by a transmitting mirror. The transmitting mirror reflects most of the light back into the ruby crystal to attain an ongoing process of stimulated emission and therefore amplification. If too much energy is coupled out of the transmitting mirror the stimulated emission will decay and die. Therefore the excitation mechanism must constantly pump energy in to the laser.

This is just a cursory look at the operation of a laser. Chapter 9 provides a more in-depth look at laser types, characteristics, properties, and operation.

Lasers in Operation   The laser has become one of the most productive devices in present-day commerce and industry. It has been used in communications, holography, medicine, direction finding, manufacturing, and many other fields. It has always been a fictional aspiration to use a laserlike death ray or an antiballistic weapon that could destroy aircraft or missiles. The advances that are being made today allow the laser to operate with more power, greater distance, and greater control. These fictional thoughts are not as far from reality as they were a decade ago.

In the communications field, lasers are utilized in audio, radio, and television transmission. In television, the scene being shot is often illuminated by laser-light scanning. Laser beams have a narrow bandwidth, high directional capability, and low divergence. These properties are extremely important in communications using modulated light. Lasers are used in production of three dimensional movies and television in a science called *holography*. Holography is also used in manufacturing processes where it is used to detect vibrations and locate strains. Holography is as new as laser technology and its future just as promising.

In medicine there is no technology that has advanced to the extent of the laser. Ruby lasers for years have been used for eye surgery. Argon ion lasers are useful as scalpels. The body skin fat and muscle are transparent to selective laser wavelengths; therefore, cancer cells can be destroyed without harming the other parts of the body. Decayed areas of the teeth can be destroyed without harm to healthy teeth.

Distance measuring for surveying and for military fire control are other uses of the laser. Laser radars are now far past their infancy.

In manufacturing, the laser is used with holography to detect strains and to measure irregular objects. Powerful lasers are used to cut stacks of cloth, drill fine holes accurately, weld, and to align objects such as aircraft wings.

In light transmission, injection laser diodes are utilized to launch light into small diameter optical fibers. Indeed, laser technology and the fiber-optic industry are moving rapidly to the forefront in all industry and especially in conjunction with the electronics industry.

2

LIGHT SOURCING, TRANSMITTING, AND RECEIVING

Light is either leaving a source or arriving at a detector or sensor. At all other times it is in motion between the two. This chapter deals with sourcing, transmitting, and receiving light.

GENERAL

The sun and the stars are the natural sources of light. The aurora is another natural source of light. Reflective sources of light include planets and moons. The amount of light energy received from the sun (solar irradiance) outside the earth's atmosphere is called the *solar constant* and is equal to 14 milliwatts per square centimeter. The atmosphere absorbs much of this. If we consider a clear day with no fog, smog, or cloud cover, the amount of solar irradiance reaching the earth is 100 milliwatts per square centimeter. The amount of light energy received from the sun is an inverse curve that is dependent primarily on the distance from the sun.

Light is the form of electromagnetic radiation that acts upon the retina of the eye and the optic nerve. This makes sight possible. The term light has been altered by technical advances to include electromagnetic wavelengths not visible to the human eye. The entire frequency spectrum that deals with wavelengths visible and adjacent wavelengths is called the *optical frequency spectrum*.

There are three basic levels of the optical frequency spectrum. These are:

1. Infrared—Band of light wavelengths that are too long for response by the human eye.
2. Visible—Band of light wavelengths that the human eye responds to.

3. Ultraviolet—Band of light wavelengths that are too short for response by the human eye.

Optoelectronics deals with this entire spectrum. You may wish to review the section called "Optoelectronic Operational Spectrum" in chapter 1 for details of the spectrum. Wavelength ranges for these areas of the spectrum are 0.005 to 0.3900 microns for the ultraviolet, 0.3900 to 0.7500 microns for the visible, and 0.7500 to 4000 microns for the infrared. These values will be repeated at significant places throughout this book.

Since light is a measureable source, there have been two scientific fields created to measure it. Photometry is involved with the visible electromagnetic wavelengths. Radiometry deals with those wavelengths that are too short or too long for the human eye to perceive. Chapter 10 explains these two scientific fields of measurement and describes techniques and applications for measurement.

The theory of the light called the quantum theory defines the photon. The photon is an uncharged particle and has an energy level of its own which is dependent on its frequency or wavelength. The higher the frequency of the energy the greater the strength of the photons.

Energy States Each atom has several energy states (see figure 2–1). The lowest energy state is the ground state. Any energy state above ground represents an excited state and can be called by names such as first or second energy states or intermediate energy states. If an atom is in one of its energy states (E4) and decays to a lower energy state (E2), the loss of energy (in electron volts) is emitted as a photon. The energy of the photon is equal to the difference between the energy of the two energy states. The process of decay from one energy state to another is called *spontaneous decay* or *spontaneous emission*.

An atom may be irradiated by some light source whose energy is equal to the difference between ground state and some energy state. This can cause an electron to change from one energy state to another by absorbing light energy. This transition from one energy level to another is called *absorption*. In making the transition the atom absorbs a packet of energy called the photon. This is similar to the process of emission.

The energy absorbed or emitted (photon) is equal to the difference between the energy states. For instance, $E2 - E1 = E_p$ (energy of photon).

$$E2 - E1 = E_p \text{ (energy of photon)}$$

also $\qquad E_p = hf$

where $\quad$ h = Planck's constant

$\qquad\qquad$ f = frequency of the radiation

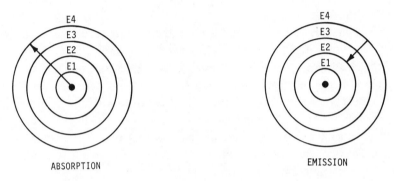

Figure 2–1 Energy States in Atoms (*Mike Anaya*)

The value of h has a dimension of energy and time. The value is 6.625×10^{-34} joule-second. The value of f is in cycles per second (hertz). Photon energy is said to be directly proportional to frequency.

Photon energy may also be expressed in terms of wavelength. You may recall that wavelength is equal to the speed of light in meters per second divided by frequency,

$$\lambda = \frac{c}{f}$$

where $\lambda =$ wavelength

$c =$ speed of light in meters per second

$f =$ frequency of the electromagnetic wave in cycles per second (hertz)

The formula can also be stated in the following manner:

$$c = f\lambda \quad \text{and} \quad f = \frac{c}{\lambda}$$

If we substitute this value of f in the formula for the energy of a photon we have a second formula that relates the energy of the photon to wavelength.

$$E_p = hf$$

$$E_p = h(\frac{c}{\lambda})$$

$$E_p = \frac{hc}{\lambda}$$

Stimulated Emission You may recall from the last discussion on energy levels (or states) that spontaneous emission occurs when an atom decays from one energy state to a lower energy state (see figure 2–2). The power that originally caused the atom to move into a higher level of energy was an electric current or a powerful light. Another way for an atom to change energy levels is to absorb a photon. This can happen only when the energy difference between two levels of an atom, the level it is in and the level to which it is being excited, are equal to the energy of the photon.

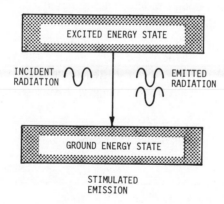

Figure 2–2 Stimulated Emission (*Mike Anaya*)

Stimulated emission is a laser process. Let's consider an atom which is in, say, a third-level energy state and will probably decay to a lower energy state. While the atom is still in the higher energy state an external source of radiation is applied that is extremely close to the frequency of the atom in the excited state. Then there is a good chance that the external source will stimulate the atom into moving from that excited state to the ground level. In the case of stimulated absorption the incident radiation moves the atom to a higher energy level. In the case of stimulated emission the atom is moved to a lower level and emits two wave packets (photons of precisely the same frequency).

The stimulated emission and the incident frequency are the same and move in the same direction and are in step (in phase) with each other. Therefore they are coherent. The time that an atom remains in an excited state is extremely short, about 10^{-8} seconds. Stimulated emission must take place when the atoms are in this excited state. Each atom has several energy states and moves between these states. The longer the time between these energy states the better the atoms are adapted for stimulated emission. Long-duration energy states are called *metastable states* and have energy state times of 10^{-6} to 10^{-3} seconds. When excitation has taken place to cause absorption, the population of the excited-state atoms exceeds the ground-state atoms; when this happens the process is called *population inversion*.

You may recall the Bohr theory of the atom from other basic atomic studies. The theory describes the atom as having a nucleus consisting of neutrons and protons. Electrons rotate around the nucleus in shells or rings. The outside ring holds valence electrons. These electrons are the farthest away from the nucleus and therefore are not held as tightly in place as the electrons closer to the nucleus. If there is enough radiant energy applied to the atom the electrons can be removed from their orbits. The electrons can then escape from the force applied by the nucleus and become free carriers. The amount of energy required to remove the electron from the valence ring (excite the valence electron into an excited state) is called the *energy gap*. Listed in table 2–1 are some of the solid-state materials used in optoelectronics along with their energy gaps.

TABLE 2–1 Optoelectronic Materials and Their Gaps

ATOM	ENERGY GAP (IN eV)
Cadmium sulfide (CdS)	2.4
Gallium phosphide (GaP)	2.2
Cadmium selenide (CdSe)	1.7
Gallium arsenide (GaAs)	1.4
Silicon (Si)	1.1
Germanium (Ge)	0.7
Lead sulfide (PbS)	0.37
Lead telluride (PbTe)	0.29
Lead selenide (PbSe)	0.26

LIGHT SOURCING

Basic Light Sources A point source is theoretically a dimensionless (no length, width, or height) point in space from which light is propagated in all directions away from the source. (See figure 2–3A.) In reality this cannot be, for any point or light source must have a finite size. Point sources can, however, appear to be a defined point such as a star that is light-years away and a laser beam. In the illustration, rays that leave point P arrive at point A in parallel lines of the direction P to A. Likewise rays leave point P to points B and C in the same manner.

In the figure 2–3B, the extended source of light is shown. This source is depicted as a point in space that is illuminated by light being radiated from several different directions. Points A, B, and C (the extended source) are illuminating point P from several directions. The point P then is illuminated by each source point. The further you move from point source (A, B, C) the light lines become more nearly parallel. From an extreme distance the extended source appears to be a point source. For example, a flashlight viewed from a distance of 5 feet would appear to be an extended source that radiated from many points on its face. That same flashlight when viewed from 500 feet would appear to be a point source.

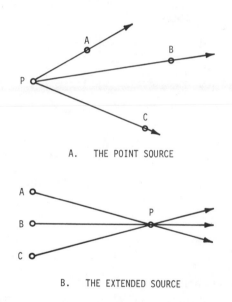

A.   THE POINT SOURCE

B.   THE EXTENDED SOURCE

Figure 2–3   Basic Light Sources (*Mike Anaya*)

Point Source Intensity   Intensity is the amount of radiant or luminous flux per unit solid angle that is diverging from the light source. Photons of light energy combine to create radiant power called *flux*. The steradian is a solid angle of a sphere equal to the area of the sphere divided by the radius squared. (See figure 2–4A.)

A.   A SOLID ANGLE

B.   POINT SOURCE INTENSITY

C.   AREA SOURCE INTENSITY

Figure 2–4 Source Intensity (*Mike Anaya*)

$$\omega = \frac{A}{r^2}$$

where ω = solid angle = 1 steradian (SR)

A = Curved surface area of the sphere

r = radius of the sphere

There are 4π steradians within a sphere. That is, there are 4π solid angles projected from a point source at the center of the sphere.

Point source intensity in radiometric terms is measured in watts per steradian. Radiant intensity (the light power of the point source) is the amount of power from a point source generated in a unit solid angle. (See figure 2–4B.)

$$I_R = \frac{P}{4\pi}$$

where I_R = Radiant intensity in watts per steradian

        P = Radiant flux in watts

        $4\pi = 4 \times 3.1416$

Point source intensity in photometric terms is measured in candelas or lumens per steradian. Luminous intensity is the amount of power from a point source generated in a solid angle. (See figure 2–4B.)

$$I_L = \frac{F}{4\pi}$$

where I_L = Luminous intensity in candelas or lumens per steradian

        F = Luminous flux in lumens

        $4\pi = 4 \times 3.1416$

Note that the solid angle is applicable to both radiometric and photometric measurements. Measurement of light intensity along with other light measurement is covered in chapter 10.

     Area Source Intensity   Intensity, once again, is the amount of radiant or luminous flux diverging from a projected source area or light source on a finite area. This flux per unit area is also called *exitance*. Area source intensity in radiometric terms is measured in watts per centimeter squared.

$$B_R = W/cm^2$$

where B_R = Radiance

        W = Watts

        cm^2 = Area of source

Area source intensity in photometric terms is measured in candelas or lumens per square centimeter.

$$B_L = L/cm^2$$

where B_L = Luminance

        L   = Lumens

        cm^2 = Area of source

     An area source whose angular distribution is a circle is called a *Lambertian source*. As you will see later on in the book, a flat LED (light-emitting diode) chip is an excellent approximation of the Lambertian area light source.

LIGHT TRANSMITTING

Electromagnetic Waves   Light is fairly predictable, tending to repeat itself in the same manner. Light energy travels in waves such as the electromagnetic waves of electricity (see figure 2–5). The waves in electricity are called *sinusoidal* or *sine waves*. This name comes from a trigonometric ratio. The sine wave has an angular rotation of 360°. It is maximum at 90° and 270°, and minimum at 0°, 180°, and 360°.

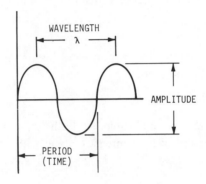

Figure 2–5   The Electromagnetic Wave (*Mike Anaya*)

The sine-wave phenomenon is the same in light waves as in electrical waves. Each wave is characterized by five basic quantities: velocity, amplitude, frequency, polarization, and wavelength. The characteristics of waves are also the same. Each wave has a positive and a negative alternation. These are identical to each other. Each light wave has a time period in which it goes through its complete cycle or alternation. Some waves take a shorter time to complete the cycle. For instance, from a home electrical socket a current of 60 cycles per second is taken. This means that during one second the cycle takes place 60 times. The time for one cycle to take place is 1/60 of a second or 16.6 milliseconds (one period).

$$T = \frac{1}{f} = \frac{1}{60} = 16.67 \text{ milliseconds}$$

where   f = Frequency in hertz

T = Time in seconds

This formula can be converted to calculate frequency when time is known.

$$f = \frac{1}{T}$$

Time for one cycle to take place is called the period measured in seconds. The number of cycles that takes place within a time period is called cycles per second or hertz.

The physical length of the wave is called its *wavelength*. It is this entity that concerns us in optoelectronics. All electromagnetic waves travel at the same velocity. That is 300,000,000 meters per second or 186,000 miles per second. The amplitude of the wave is the magnitude of the wave vibration. The frequency of the wave is the number of waves passing a point in a second. The wavelength is measured in meters per second and is called lambda (λ). The wavelength is the distance between two consecutive wave peaks. Calculations of wavelength are as follows:

$$\lambda = \frac{v}{f}$$

where λ = Wavelength in meters per second

v = Velocity of light in meters

f = Frequency of the wave in hertz (cycles per second)

The optical spectrum does not only deal with frequencies and wavelengths. It deals with color. Each wavelength is a color. The human eye sees red on one side of the frequency spectrum and violet on the opposite side of the frequency spectrum. Center frequencies are blue, green, yellow, and orange. White light such as that from the sun or a flashbulb can be separated or dispersed into all the colors of the frequency spectrum. Rays of light are propagated from one of three basic sources: panchromatic light is radiation that includes all the wavelengths within the spectrum; heterochromatic light is light that has several very distinct wavelengths; monochromatic light has one wavelength or an extremely narrow band of very distinct wavelengths, such as monochromatic laser beams.

Polarization of Waves   Light waves are either polarized or unpolarized. If the waves oscillate randomly through space or material and have no general direction the waves are said to be unpolarized. If the waves oscillate at the same amplitude and trace out a circle in a plane perpendicular to the direction of propagation, the waves are said to be circularly polarized. If the waves oscillate at the same amplitude along a line in a plane perpendicular to the direction of propagation, the waves are said to be linearly or plane polarized. There are other polarizations such as the elliptically polarized wave. (See figure 2–6).

A beam of unpolarized light may be the result of two beams combined in two different planes at two different amplitudes with no phase relationship. The directional variations of the two waves are not related nor are their magnitudes.

It is extremely important that waves be polarized in optoelectronics.

Figure 2–6 Polarization of Waves (*Mike Anaya*)

Within optoelectronic systems such as the laser, linear polarizers are used to separate out all rays that are not parallel to the propagation axis. The beam can be tested for polarization by rotating the polarizer by 90°. This should cancel the beam by orienting the polarizer so that its axis is perpendicular to the plane of the transmitted beam. Chapter 6 illustrates and describes the optic polarizer.

    Coherence Light sources such as fluorescent lamps do not radiate coherently. The light from these lamps is radiated in many frequencies and just as many phases. As a result the light beams do not have direction, certainly are not polarized, and act in an extremely unpredictable manner. This incoherency can be compared with the light of a laser beam, which is directional and polarized. Its wave fronts are defined and its frequency along with phase relationships are constant. In effect the beam is monochromatic (one color) and coherent.

    The coherency of the light waves is imperative to good laser operation. In order to have coherency the waves must have a consistent relationship between their wave troughs and wave crests. Wave troughs are points of minimum vibration and wave crests are points of maximum vibration. Coherency is divided into two major relationships, spatial and temporal. To be spatially coherent, waves must have phase correlation across a wave front at any point in time. The spatially coherent waves also maintain their shape in relation to time. To be temporally coherent, waves must be frequency constant so that phase correlation is in the direction of propagation. The phenomenon of coherence allows the laser beam to be focused to a minute unit area. A conventional source cannot be focused without loss of power. Furthermore, coherent light can be made to produce constructive intereference of waves such as those used in holography.

Figure 2–7 Coherence (*Mike Anaya*)

Beam Divergence   The light beam in a laser is coherent. It can therefore be focused to a minute beam. The laser beam will tend to fan out or diverge as it travels away from its power source. The beam may have many different cross section TEM shapes (transverse electromagnetic modes). However, for purposes of explanation, we shall assume that the beam is round (TEM_{00}).

In figure 2–8 a typical beam is illustrated. D1 represents the diameter of the beam at its aperture (output from the laser). D2 represents the diameter of the beam at the detector, receiver or somewhere in space. L is the length of the beam at that point. Angle θ is the angle of divergence or fanning out. The angle in the illustration is extremely exaggerated for the divergence of a laser beam is a very narrow cone. The calculation for the diameter of beam divergence is as follows:

$$D2 = L \angle \theta + D1 \text{ (for very small angles)}$$

where   D2 = Diameter of beam divergence

L = Length of projected beam

D1 = Diameter of the projection at the aperture

$\angle \theta$ = Angle of divergence (in radians)

The angle θ must be converted to radian measure for these calculations. You may recall from your geometric studies that a radian is equal to 57.3°. Two pi radians ($2\pi R$) are equal to 360° angular rotation about a point. Since the angle of divergence is quite small, the radian measure of that angle is usually in milli-

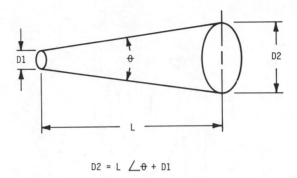

$$D2 = L \angle\theta + D1$$

Figure 2–8 Beam Divergence (*Mike Anaya*)

radians. One milliradian is one-thousandth of a radian or 0.0573° which represents 3.44 minutes of arc.

Often the beam divergence is specified in half-angle divergence. In this case the half-angle would be θ divided by 2 with minor considerations given to the aperture diameter. Typical laser beam divergence is 1 milliradian for gas lasers and 10 milliradians for dye lasers. Semiconductor lasers have large beam divergence angles such as 500 milliradians.

Since it is imperative that the beam be thin with low divergence angles, it behooves the designer to plan for focusing to preclude large divergence angles. Focusing is accomplished with lenses. The ability of a lens to focus is dependent on its focal length and the beam source divergence angle. There are limits to which the beams can be focused. These limits are approximately equal to the square of the beam wavelength. For large apertures, lenses may not be needed.

LIGHT RECEPTION

General The reception of light is completely dependent on what use or application is intended. If the desire is to light a room, the light should reflect diffusely from most objects in the room with absorption dependent on the materials and colors in the room. There is little or no use for refractive light. If the light were to be used for fiber-optic applications, fiber ends would reflect at zero percent if possible and the light refracted into the fiber would be maximum. The amount of absorption within the fiber would be held to a minimum. If the light were to be used for photodetection the amount of absorption would be extremely high (100 percent if possible). The amount of reflection would be held to a minimum. Thus the application decides the use of materials that are low or high absorbers, reflect or do not reflect, and refract or do not refract.

Interference Effects   The study of interference and its peculiarities is an extremely broad field. The field does have applications in optoelectronics, primarily in laser technology and holography. Interference effects may be produced by coherent light. Broadly defined, interference is the interaction between two beams of light. The two beams should be coherent for the interaction cannot take place with incoherent light. The interaction produces interference fringes. The fringes are variations of resultant amplitudes of the interaction as the waves vary from point to point in a field of view. The variations are due to path differences where the waves reinforce each other at certain points and oppose each other at other points. These interference patterns allow the relative coherency of a light source. The measurement is made by analysis of the interference fringe pattern.

The first attempt to observe interference patterns was made by Francesco Grimaldi in the seventeenth century. He placed a monochromatic light from a single source in front of a screen with two slits in it. He did observe interference but it was not clear. It was later shown by Thomas Young that the interference could only be seen clearly if the light source is small. (See figure 2–9.) Young's experiment involved a small light source radiated through a slit in a screen which he called a source slit. A second screen with two slits, each equidistant from a horizontal line extending from the source slit was placed between the source and the viewing screen. Some waves from the source are intercepted by slit 1. Slits 2 and 3 sample these waves at two points. If the phase relationship of the two wave fronts is fixed, the interference pattern will appear on the screen as bright and dark lines or fringes. The fringes may be analyzed thusly. If the path difference (D) between the two path lengths is equal to half a wavelength, then the waves arrive at point P, 180° out of phase with each other. Similarly, if the path difference (D) between the two paths is equal to a full wavelength, then the waves arrive at point P in phase.

Figure 2–9 Interference Effects (Young's Experiment) (*Mike Anaya*)

If the interference patterns projected on the screen are dark fringes, then the two waves are out of phase. Out-of-phase interference patterns are called *destructive interference*. In figure 2–10, interference patterns 180° out of phase are illustrated. The wave pattern is a standing wave such as is found in laser cavities. However, if the interference patterns projected on the screen are bright fringes then the two waves are in phase. In-phase interference patterns are called *constructive interference*. In figure 2–10B, these in-phase waves are summed to a higher amplitude resultant.

Incoherent light does not produce spatial or temporal coherence. The wave front is not clearly defined and the phase relationships are constantly shifting; therefore, incoherent light sources cannot be used for the formation of interference fringes. However, if the incoherent light source is projected through a narrow-band filter it may be utilized. The filter tends to isolate monochromatic waves of the light. That light may contain some degree of temporal coherence.

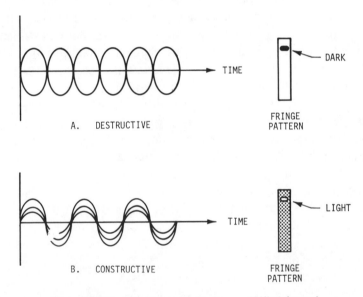

Figure 2–10 Interference Patterns (*Mike Anaya*)

Incidence Incidence is the amount of flux per unit area that is normal or perpendicular to a surface or detector. If the flux is not normal (not perpendicular), then the normal component of the angular flux is the incidence. In radiometric terms, incidence is called *radiant incidence* or *irradiance*. Irradiance (E_e) is measured in watts per square meter ($E_e = W/m^2$). Incidence in photometric terms is called *luminous incidence* or illuminance (also called illumination). Illuminance is measured in lux (lx) or lumens per square meter (lm/m²).

When light falls on a surface it is partly reflected from the surface and partly absorbed. It may also be converted by the surface into some other form of energy. In this case it becomes part of an active surface such as a photodiode or phototransistor. In any event, incidence is how much flux falls on the surface area. From a point source, the law of illumination of light falling on a given area is called the *inverse square law*. (See figure 2–11.) You will note that one lumen of flux illuminates 1 square foot of area from 1 foot away at 1 footcandle (1 lumen/square foot). At 2 feet from the point source, an area of 4 square feet is illuminated to ¼ footcandle. Finally, at 3 feet from the point source, an area of 9 square feet is illuminated to 1/9 footcandle. The distance away from the source is squared to provide the illuminated area. The illumination of that area is the inverse to the square of the distance.

Figure 2–11 Point Source Illumination (The Inverse Square Law) (*Courtesy United Detector Technology, Inc.*)

Illumination from an area source must be looked at from a different perspective. The area source must be assumed to be an extended source. That is, each point or element on the surface contributes to the luminous intensity of the source. The luminous intensity in a given direction, divided by the projected area of the source is called the *luminance* of the source. In the figure 2–12, area source illumination is illustrated. An angle is made between a normal line (perpendicular to the surface) and the incident flux lines. The angle is used to calculate the

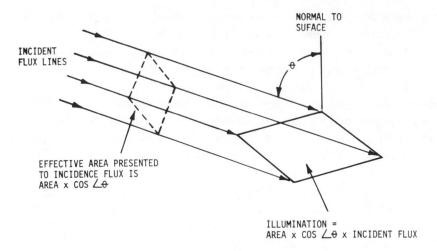

Figure 2–12 Area Source Illumination (The Cosine Law) (*Mike Anaya*)

effective area presented to incident flux. The effective area is equal to area times cosine $\angle\theta$. The illumination in lumens per square feet is equal to the area of the receiver times cosine $\angle\theta$ times the incident flux. Please refer to chapter 10 for more details on photometry/radiometry.

Reflection When light beams or rays strike a surface, they reflect from that surface either in a diffuse manner or a specular manner or in some cases both. The amount of reflection is largely due to the type of material on the surface of the reflector. (See figure 2–13.)

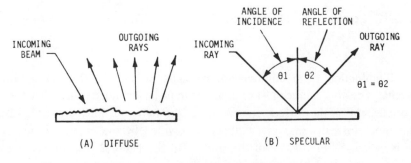

Figure 2–13 Reflection (*Mike Anaya*)

Diffuse Reflection   Diffuse reflectors scatter the light outward in several beams depending, of course, on the surface material and the amount of light the material absorbs. Virtually all material absorbs some light. In figure 2–13A, the principle of diffuse reflection is illustrated. A smooth surface such as a freshly painted house would diffuse light in rather equal and alike reflected rays. Another surface such as a rock-covered roof would cause light rays to be reflected in scattered, nondescript lines.

Surface or Specular Reflection   Whenever a beam of light from one medium such as air, strikes a second mirror-like medium such as glass, part of the beam is reflected and the other part refracted. (See figure 2–13B.) The angle that the incident rays strike the second medium is called the *angle of incidence*. This angle is the angle made by the incident beam and a line normal (perpendicular) to the boundary of the two media. Part of the incident beam is reflected at an angle called the *angle of reflection*. This angle is made by the reflected beam and a line normal (perpendicular) to the boundary of the two media.

Angle of incidence = Angle of reflection

If the angle of incidence varies, so does the angle of reflection by the same amount. The angle of incidence, the angle of reflection, and the angle of refraction all lie in the same plane.

Fresnel Reflection   Fresnel reflections become Fresnel reflection losses when dealing with fiber optics. Fresnel losses are losses that are a result of the differences between refractive indexes of the glass fiber core and its cladding. Variations of Fresnel reflections are related to unpolarized light, light polarized perpendicular to the plane of incidence, and light polarized parallel to the plane of incidence.

Surface Refraction   Whenever a beam of light passes from one medium such as air to another medium such as water, the beam separates at the intersection. Part of the beam is reflected back into the incident medium (air) and part is refracted into the second medium (water). The angle made by the incident beam and a line normal (perpendicular) to the intersection is called the angle of incidence. The angle made by the reflected beam and the line normal to the intersection is called the angle of reflection. The angle of incidence and the angle of reflection are equal. The angle made by the refracted beam and a line normal (perpendicular) to the intersection is called the angle of refraction. (See figure 2–14.)

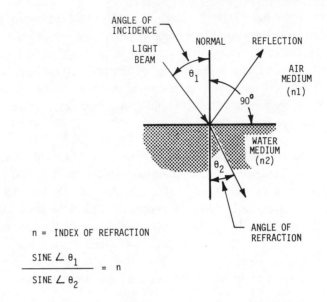

Figure 2–14 Surface Refraction (*Mike Anaya*)

        The ratio of the indexes of refraction of the two indexes of refraction is thus:

$$n = \frac{n2}{n1}$$

or   n = index of refraction.

The ratio of the sine of the angle of incidence to the sine of the angle of refraction is equal to the index of refraction.

$$n = \frac{\text{sine } \angle \theta 1}{\text{sine } \angle \theta 2}$$

therefore

$$\frac{n2}{n1} = \frac{\text{sine } \angle \theta 1}{\text{sine } \angle \theta 2}$$

or   n1 sine $\angle \theta 1$ = n2 sine $\angle \theta 2$

Furthermore, the ratio is a constant ratio. Whenever the angle of incidence changes the angle of refraction changes to retain the ratio. The law is Snell's law. Snell was a Dutch astronomer and professor of mathematics at the University of Leyden in Holland. In fiber-optic applications it is imperative that the light incident to the fiber end be refracted into the fiber core. The amount of light refracted is determined by the refractive indices of the air and the fiber core. Refer to chapter 6 for fiber-optic transmission angles.

Scattering   Scattering is differentiated from absorption in the following manner. With true absorption, the intensity of the beam is decreased in calculable terms as it penetrates the medium. Light energy absorbed in the material is converted to heat motion of molecules. Consider a long tunnel where you could only see light from one end. As you walked nearer that end the light gets brighter. The light is absorbed as it travels through the tunnel at the absorption rate of the air medium. If the tunnel were then filled with a light cloud of smoke, the smoke would scatter some of the light from the main beams, therefore the intensity of the light from a fixed distance would decrease. You may observe scattering effects by watching dust particles as the sun shines in a window. Parts of the rays are scattered by the dust particles.

A scattering type known as *Rayleigh scattering* is caused by micro irregularities in the medium. A wave passing through the medium strikes these micro irregularities in the mainstream of the waves. The waves reflected from the microparticles are spherical and do not follow the main wave, but scatter. Therefore, the intensity of the beam is diminished. In fiber optic application the composition of the glass must be considered to ensure low Rayleigh scattering. Silica has low scattering losses.

Dispersion   Whenever a beam of light enters matter its velocity decreases within that medium. The subject of dispersion deals with the speed of light in a medium and the variation in conjunction with wavelength. The speed of light in a medium can be calculated by its refractive index. The refractive index (n) of any material is the ratio of the speed of light in a vacuum to its speed in the medium.

$$n = \frac{c}{v}$$

where   n = Refractive index of the medium

c = Speed of light in a vacuum

v = Speed of light in a medium

Then the speed of light (v) in the medium may be calculated by

$$v = \frac{c}{n}$$

The refractive index for air is 1.000, whereas optical glasses have values of 1.520 to 1.720 for refractive indexes. Any change then in refractive index would represent a change in velocity.

The dispersion of each material is different. Curves are supplied by manufacturers that plot wavelength against refractive index. As the wavelength decreases the refractive index increases and the dispersion ($dn/d\lambda$) increases. The rate of increase becomes greater at shorter wavelengths. Dispersion is the spreading of light rays. In the propagation of light rays through a medium such as an optical fiber the problems of dispersion are multiplied. With fiber optics there are two types of dispersion, intermodal and intramodal. Intermodal dispersion is caused by the propagation of rays of the same wavelength along different paths through the fiber medium. This results in the wavelengths arriving at the opposite end of the fiber at different times. Intramodal dispersion is due to variations of the refractive index of the material that the fiber was made from. Dispersion is measured in nanoseconds per kilometer (ns/km) in fiber-optic applications.

Absorption   Whenever a beam of light enters matter its intensity decreases as it travels further into the medium. There are two types of absorption, general and selective. General absorption is said to reduce all wavelengths of the light by the same amount. There are no substances which absorb all wavelengths equally. Some material such as lampblack absorbs near 100 percent of light wavelengths. With selective absorption the material selectively absorbs certain wavelengths and rejects others. Almost all colored things such as flowers and leaves owe their color to selective absorption. Light rays penetrate the surface of the substance. Selective wavelengths (therefore colors) are absorbed, whereas others are reflected or scattered and escape from the surface. These wavelengths appear in color to the human eye. In optoelectronics it is imperative that near 100 percent of incident beams be absorbed in devices such as solar cells and detectors. On the reverse side, zero percent absorption is important in the transmission of light waves in an optical fiber cable and lenses.

Photodetection   Photodetection includes three phenomena: photoemission, photoconduction, and photovoltaic action. Photoemission is defined as the emission of electrons by light energy. The effect of light on certain materials causes electrons to be displaced from the surface. The emitted (radiated) electrons are called photoelectrons. Light is applied to the cathode of a basic phototube. The light causes photoelectrons to be drawn through a load by the positive anode. Thus the light determines the current flow and the voltage drop across the load resistance. Current flow is also controlled by the potential applied to the anode.

Operation through photoconduction involves a change in resistance of the photosensitive material. A wafer of photoconductive material is placed underneath a glass window to protect it from exposure. The photoconductive material is tied to a load resistance and a power source. The clear glass window allows light radiation to strike the photoconductive material freeing valence electrons. The resistance of the photoconductive material decreases, causing current through the load to increase.

The photovoltaic cell uses dissimilar metals to generate an electromotive force in response to radiated light. A light-sensitive material is placed beneath a thin layer of transparent metal and next to a dissimilar metal. The light-sensitive material is exposed to radiation through the thin transparent metal that acts as a filter. When exposed, free electrons are removed from the light-sensitive material, causing electrons to flow to the dissimilar metal. This creates current flow and a difference of potential between the two terminals connected to the load. Photodetection is described in detail in chapter 3.

Electroluminescence The most modern of the optoelectronic effects is electroluminescence. This effect involves the application of electrical energy to a semiconductor material causing it to emit light (radiated) energy. Current is caused to flow through a load from a negative terminal through negatively charged metal. Electrons are released and are drawn through positively charged metal to the positive terminal. An energy gap is developed between the negatively charged metal and the positively charged metal at their junction. It is at this junction that electrons combine with positive charges to create photons. These photons cause the device to radiate light through a glass envelope similar to a small lightbulb.

Combinations of the two metals used provide different light wavelengths. Material such as gallium arsenide added to silicon provides one wavelength. Adding phosphorus to the combination provides another wavelength. Different materials will cause radiated light to appear to the human eye as different colors, such as green or red. Other materials will radiate invisible light in the ultraviolet or infrared ranges of the spectrum. This is an oversimplification of the process. Much more detail will be provided in chapter 4.

3

PHOTODETECTION

The human eye detects visible radiation in wavelengths between 3900 to 7500 angstroms. Other detectors are capable of detecting electromagnetic radiation from radio waves to x-rays. The wavelength detection that we are considering includes ultraviolet, visual, and infrared wavelengths, that is, wavelengths from 0.005 microns to 4000 microns. There are three basic forms of photodetection. They involve photoemission, photoconduction, and photovoltaic action. All quantum detectors respond directly to the action of incident light waves. The first, photo-emission, involves incident light that frees electrons from a detector's surface. This usually occurs in a vacuum tube. We shall describe this action in the next paragraph. With photoconduction the incident light on a photosensitive material causes the photodetector to alter its conduction. The third, photovoltaic action, generates a voltage when light strikes the sensitive material in the photodetector. In application, solid-state detectors operate in either the photoconductive mode or the photovoltaic mode.

General

Photoemission Photoemissive action usually takes place in a vacuum tube. The action is due to radiant energy striking a sensitive material causing an emission of electrons. The effect of light on certain materials causes electrons to be displaced from the surface. The emitted (radiated) electrons are called photoelectrons. In figure 3–1, light is applied to the cathode of a basic phototube. The light causes photoelectrons to be drawn through a load by the positive anode. Thus the light

NOTE: CATHODE IS COATED WITH RADIATION
SENSITIVE METAL. ELECTRONS ARE
RELEASED AS RADIATION IS APPLIED.

Figure 3–1 Photoemission (*Mike Anaya*)

determines the current flow and the voltage drop across the load resistance. Current flow is also controlled by the potential applied to the anode.

When light is applied to the surface of the cathode, it illuminates the cathode. Illumination is expressed in lumens or footcandles. One lumen is the amount of luminous flux of one square foot of a spherical surface, one foot from a one-candle source. A footcandle is the intensity of illumination at all points on the illuminated surface one foot from a one-candle source. (Stated in comparison, intensity of illumination is one footcandle when the amount of luminous flux is one lumen per square foot.)

Alkali metals have the best ability to emit electrons because of light. Some of these metals are cesium, lithium, and rubidium. Thin coats of the alkali metal are placed on the cathode to provide a radiation-sensitive coating. Some phototube devices are the gas-filled phototube, the multiplier phototube, and the vacuum phototube. Phototubes are used in applications such as relay control and film/sound reproduction.

This is the extent of coverage of photoemission in this book. The book is oriented to solid-state devices, therefore vacuum tubes do not fall within its scope. You may wish to refer to vendor application notes and general catalogs if you wish to pursue photoemissive action.

Photoconduction   Operation through photoconduction involves a change in resistance of the photosensitive material. A wafer of photoconductive material is placed underneath a glass window to protect it from exposure. The photoconductive material is tied to a load resistance and a power source. The clear glass window allows light radiation to strike the photoconductive material freeing valence electrons. The resistance of the photoconductive material decreases, causing current through the load to increase. The resistance of the photoconductive material may change from several million ohms to several hundred depending on the current demand of the device. (See figure 3–2.)

NOTE: PHOTOCONDUCTIVE MATERIAL DECREASES
              RESISTANCE AS RADIATION IS APPLIED.

Figure 3–2 Photoconduction (*Mike Anaya*)

Photovoltaic Cell Operation   Operation of the photovoltaic cell involves the use of dissimilar metals to generate an electromotive force in response to radiated light. In the figure 3–3, a light-sensitive material is placed beneath a thin layer of transparent metal and next to a dissimilar metal. The light-sensitive material is exposed to radiation through the thin transparent metal which acts as a filter. When exposed, free electrons are removed from the light-sensitive material, causing electrons to flow to the dissimilar metal. This creates current flow and a difference of potential between the two terminals connected to the load.

Photodetector Types   The most basic form of detection is the photoresistor. The photoresistor is a small slice of photoconductive material whose resistance increases or decreases as light energy is applied. Electrons are released by the light and flow toward a positive power supply. The basic assumption in the photoresistor is to control electrical energy from the power supply. The photo-

Figure 3–3 Photovoltaic Cell Operation (*Mike Anaya*)

resistive material is nonreflective. There are no junctions in a bulk photoresistor, therefore lead polarity doesn't matter.

A second detector is the single-junction photodiode. A photodiode is the optical version of the standard diode. It is constructed of a PN junction. Photons of light energy are absorbed into the device. Hole-electron pairs are generated. The pairs are combined at different depths within the diode depending on the energy level of the photon. A wide, thin surface area is used to ensure maximum absorption. Current flow is dependent on the amount of radiation absorbed. Photodiodes operate in the photoconductive mode with dc bias applied. They operate in the photovoltaic mode without bias.

The solar cell is a photodiode that is heavily doped. The depletion area is extremely thin. The cell is coated to avoid reflection. Hole-electron pairs diffuse to the depletion area of the diode where they are drawn out as useful current. Output current is dependent on input radiation. Solar cells are not biased and are photovoltaic in operation.

Phototransistors are two-junction devices that have a large base area. The base region of the phototransistor absorbs the photons of energy and generates hole-electron pairs in the large base-collector region. The collector, being reverse-biased, draws the holes toward the base and the electrons toward the collector. The forward-biased base-emitter junction causes holes to flow from base to emitter and electrons from emitter to base. Forward bias causes the phototransistor to operate just as the conventional transistor operates. The basic function, then, is that the light energy induces the transistor to operate. Collector bias is provided by a power supply. Base bias is provided by the incident light. This in turn controls collector current.

Other detection devices include photo field-effect transistors, the photothyristor, and opto-isolators. Photo field-effect transistors are constructed in the same manner as the standard-junction FET. A lens is used to focus light on the gate. The gate, being light-sensitive, excites electrons from the gate into a conduction band. Electron movement causes current flow and a voltage drop. The voltage drop induces the FET load current through its drain resistance and thus modifies the drain to source voltage. Photoposition detection devices provide an output that is proportional to the direction of the input beam. The photothyristor is a four-layer PNPN device. Photons of energy create hole-electron pairs in a thin junction. These pairs are drawn across all three junctions producing a current from anode to cathode. The opto-isolator combines a light source with a detector within a single package. Current is applied to a source LED (light-emitting diode). The LED emits photons that are detected by a detector. The detector can be in the form of a photoresistor, photodiode, phototransistor, or photothyristor.

Each of the devices and detection methods that we have discussed has applications in the infrared and the ultraviolet wavelengths with detector-material changes. Typical detection devices are described and illustrated in chapter 6.

Photodetection Materials   A detector material is specially selected so that when it is exposed to light rays it will absorb the light energy. If the detector responds to the light energy (flux), and not the wavelength, the detector is said to be nonselective. If the detector responds by varying detection at different wavelengths, it is selective. Selectivity or responsivity is the detector's response per unit of light. Wavelength responsivity is called *spectral response.* Frequency response is the speed by which the detector responds to changes in radiation amplitude. Fluctuations in output current and/or voltage are referred to as *noise.* Noise is usually caused by current that flows in the detector regardless of whether light is applied. Current, such as this, is termed *dark current,* because it flows even without radiation. A common specification for a detector is the signal-to-noise ratio. This ratio is the noise current divided by the signal current.

Photoresistor resistor materials are called *bulk photoconductors.* Bulk photoconductors are usually made from cadmium sulfide (CdS), cadmium selenide (CdSe), lead sulfide (PbS), and silicon (Si). The type of material is dependent on the application. Bulk photoconductors do not have a junction, they are simply made from one material. This material should respond to a broad range of wavelengths from ultraviolet through infrared. The ultimate purpose of the bulk photoconductor is to convert light intensity to current flow, therefore the material selected should free electrons easily with the application of light. Light should not reflect but be transmitted easily into the detector. To be efficient the bulk (resistive) material should be extremely sensitive to light radiation and be as easy to apply as a resistor. The primary problem with photoresistive material is that a problem arises with temperature stability for faster-acting materials.

The photodiode is usually made from silicon or germanium much the same as standard diodes. There are basically two processes in which the photodiode material is formed. These are the planar process and the Schottky barrier process. The planar process uses silicon as the solid and gas as dopants. By diffusion, the dopant gases penetrate the solid surface of the silicon. Diffusions may be made on diffusions, therefore several layers of dopants may be diffused on one device. The Schottky barrier photodiode operates in much the same manner as the planar diffuse. Their construction differs in that the Schottky barrier diode has its active region metallized to an N-type bulk silicon material. The Schottky barrier photodiode has some advantages over the conventional photodiode in that it is easier to manufacture and operates at wavelengths of less than 500 nanometers. The conventional photodiodes operate well at wavelengths over 500 nanometers. The reason for the wavelength difference is that short wavelengths are absorbed near the surface of the diodes while long wavelengths may penetrate deeper into the structure. Again, as in bulk resistors, temperature is a continuous problem. The PIN diode (another photodiode) has a heavily doped N and P region with a very lightly doped I (intrinsic) region between. The resistance of the I region is from 10 ohms per cubic centimeter, while the N and P regions are less than 1 ohm per cubic centimeter. The intrinsic layer provides a large depletion area that allows the wavelengths to penetrate deeper into the diode. This large depletion area complements faster speeds, lower noise, and greater efficiency at longer wavelengths.

The phototransistor is made from silicon or germanium and has much the same characteristics as the standard or conventional transistor. Phototransistors are constructed in much the same manner as photodiodes and standard transistors using the planar diffused method. The base area of the phototransistor is usually made large so as to provide an area which incident light can penetrate and generate electron-hole pairs. Phototransistors are subject to the typical problem of all transistors, temperature variations. These problems may be solved with biasing techniques and thermal-stability resistors.

Infrared detectors must use thermal detectors because their wavelengths are longer. Lead sulphide (PbS) responds to infrared wavelengths of 2 to 4 microns. Indium antimonide (InSb) is sensitive to infrared wavelengths up to 5.6 microns. Silicon (Si) and germanium (Ge) are only sensitive to the near-infrared wavelengths.

Ultraviolet detectors utilize materials such as telluride, cesium, or a fluorescent coating of sodium salicylate as cathodes within phototubes. Bulk semiconductors made of cadmium selenide (CdSe) may respond to ultraviolet in the range of 300 to 400 nanometers.

PERFORMANCE CURVES FOR PHOTODIODE DETECTORS

Figure 3–4 is a set of curves for the United Detector Inc. models PIN–LSC/5D and PIN–LSC/30D diode detectors.

Spectral Response Curve   The first curve in the illustration is the spectral response curve on the top left side. The curve plots wavelength in nanometers against relative sensitivity. Note that the peak relative sensitivity occurs around 850 nanometers. The spectral range is between 350 and 1100 nanometers.

Spectral Response versus Temperature   This curve is on the top right in the illustration. There are four curves involved here at four separate temperatures in centigrade. The curves plot wavelength in nanometers against percent of responsivity variation. Note that at all temperatures zero variation occurs between 500 and 800 nanometers with falloff both positive and negative at other wavelengths.

Source Resistance versus Temperature   The curves for two photodiode detectors on the bottom left of the illustration plot resistance in ohms against temperature in degrees centigrade. The curves are logarithmic and show an operating range at temperatures between 0 and +75°C.

Capacitance versus Bias Voltage   The curve on the lower center of the illustration plots bias voltage in volts against capacitance in picofarads. The two curves are logarithmic. They show varying capacitance at maximum bias levels. A typical capacitance value would be 110 picofarads at 10-volts bias on the PIN–LSC/5D curve.

Dark Current versus Temperature   These curves are on the lower right side of the illustration. They plot temperature in centigrade against dark current in microamperes. The two curves are logarithmic. They show varying dark current at varying temperatures. Dark current you may recall is that current which flows regardless of the devices exposure to radiation. A typical dark current value would be 0.003 microamperes with 5-volts bias at 23°C ambient temperature.

Figure 3–4 Performance Curves for Photodiode Detectors (Courtesy United Detector Technology, Inc.)

Other Curves There are other curves for photodiodes worth mentioning. The linearity curve plots incident radiant energy in watts against photocurrent in amperes. The current versus voltage characteristics (I–V) plot bias voltage against photocurrent.

PERFORMANCE CURVES FOR PHOTOTRANSISTORS

Figure 3–5 is a set of performance curves for phototransistors. This particular set is for the Litronix LPT–100 series phototransistor. Other phototransistors have similar curves.

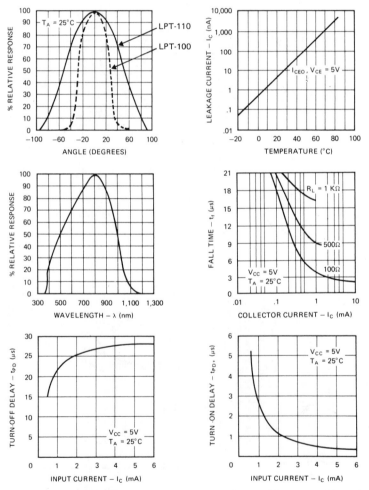

Figure 3–5 Performance Curves for Phototransistors (*Courtesy Litronix, Inc.*)

Angular Response   The curve on the top left of the illustration plots angle in degrees against relative response in percent. Note that the peak response is at an angle of 0°. The LPT–100 has a spread of around −30 to +30°. The LPT–110 has a wider spread.

Collector Dark Current versus Temperature   The center top curve plots temperature in centigrade against leakage current in nanoamperes.

Spectral Characteristics   The spectral curve on the top right of the illustration is probably the curve most frequently provided by manufacturers for photoelectric devices. This curve as you can see peaks at 800 nanometers with a spread of around 500 to 1000 nanometers.

Fall Time versus Collector Current   The curve on the bottom left is a logarithmic curve which plots collector current in amperes against fall time in microseconds, with several loads (R_L). Fall time is defined as the time required for the collector emitter current (I_{CE}) to decrease from 90 to 10 percent of peak value. Rise time is defined as the time required for collector emitter current to increase from 10 to 90 percent of peak value.

Turn-Off Delay Times for Current   The curve in the lower center of the illustration plots input current in milliamperes against turn-off delay in microseconds as current increases to 2 milliamperes, then tends to level off.

Turn-On Delay Times   The lower right curve plots input current in milliamperes against turn-on delay in microseconds. The curve drops dramatically from 5 to 1 microseconds as current increases to 2 milliamperes, then tends to level off.

Other Curves   There are other characteristic curves for phototransistors worth mentioning. These are photocurrent characteristics, collector current versus collector voltage and collector base characteristics. The latter two are also standard transistor curves.

SPECIFICATIONS OF PHOTODIODE DETECTORS

Table 3–1 is a rather standard set of specifications for a photodiode detector. This particular set establishes maximum and minimum values for the United Detector Technology Inc. models PIN–LSC/5D and PIN–LSC/30D. Other photodiodes have similar specifications. Spectral range, responsivity, dark current, and breakdown voltages are standard specifications provided for all photodiodes. Rise and fall times are equally standard parameters.

TABLE 3–1   Specifications for PIN-LSC/5D and PIN LSC/30D photodiodes

PARAMETERS AND (UNITS)		PIN-LSC/5D			PIN-LSC/30D		
		MIN	TYP	MAX	MIN	TYP	MAX
Recommended Mode of Operation			Photovoltaic			Photovoltaic	
Spectral Range @ 5% of Peak (nm)			350–1100			350–1100	
Responsivity at Peak λ (amps/watt)		–	0.55	–	–	0.55	–
Uniformity of Responsivity (with 1 mm spot dia)		–	±2%	±7%	–	±2%	±7%
Position Sensitivity at Peak λ (amps/watt/cm)		–	2.2	–	–	0.36	–
Position	Central 25%	–	0.2%	0.6%	–	0.2%	0.6%
Linearity	Central 75%	–	1.0%	5.0%	–	1.0%	5.0%
Null Point Drift (μm/°C)		–	2.0	10.0	–	3.0	10.0
NEP @ Peak λ 1 kHz (watts/Hz$^{1/2}$)		–	2×10^{-13}	–	–	5×10^{-13}	–
Dark Current (μa)	@ 5V Bias	–	0.003	0.05	–	0.1	5.0
Breakdown Voltage @ 100 μa (volts)		50	200	–	5	–	–
Source Resistance (MΩ)		10	100	–	1	10	–
Series Resistance (kΩ)		–	10	20	–	2	10
Capacitance (pF)	@ 0V Bias	–	190	380	–	14700	29000
	@ 5V Bias	–	77.5	155	–	6200	9300
	@ 10V Bias	–	55	110	–	–	–
	@ 50V Bias	–	24	48	–	–	–
Rise Time 10%–90% (μs)	@ 5V Bias	–	0.5	–	–	7.0	–
Fall Time 90%–10% (μs)	@ 5V Bias	–	0.5	–	–	7.0	–
Recommended Maximum Frequency Usage (kHz)	@ 5V Bias	–	–	10	–	–	2
Output for 10% Linearity (ma)	@ 0V Bias	0.01	0.02	–	0.05	0.1	–
	@ 5V Bias	0.1	0.2	–	0.5	1.0	–
Approximate Saturation Level (mw) @ 5V Bias		–	2	–	–	10	–

Courtesy United Detector Technology

SPECIFICATIONS OF PHOTOTRANSISTOR DETECTORS

Table 3–2 is a standard set of specifications for a phototransistor detector. This particular set is for the Litronix LPT–10C series transistor. Other phototransistors have similar specifications. Photocurrent, acceptance angles, collector-emitter reverse voltage, and collector-emitter cutoff current are standard specifications provided for all phototransistors. Temperature, power dissipation, and maximum current/voltage are also standard specifications.

TABLE 3–2 Specifications for Litronix LPT-100 phototransistor.

PHOTO CURRENT AT $V_{CE} = 5V$ $H = 5$ mW/CM² (mA)				* ACCEPTANCE ANGLE $I_c = .51_c$ ON AXIS ± (°)	COLLECTOR-EMITTER REVERSE VOLT V_{CEO} MAX V	COLLECTOR-EMITTER CUTOFF CURRENT $V_{CE} = 5V$ I_{CEO} (nA)		PACKAGE DIMENSIONS	
TUNGSTEN			GaAs			TYP	MAX	PAGE	DWG
MIN	TYP	MAX	MIN						
0.2	1.4		.6						
1.0		3.0	.6	30°				32	34
1.3		2.6	.6		30	2.0	100		
0.2	.88		.6						
0.6		1.8	.6	50°				32	35
0.8		1.6	.6						

Maximum Temperatures/Humidity
  Storage Temperature .... −55°C to +100°C
  Operating Junction Temperature .... −55°C to +85°C
  Relative Humidity at Temperature .... 98% at +65°C
Maximum Power Dissipation
  Total Dissipation at +25°C Case Temperature .... 200 mW
  Total Dissipation at +25°C Ambient Temperature .... 100 mW
Maximum Voltages
  BV_{CBO} Collector to Base Voltage 50V
  LV_{CEO} Collector to Emitter Sustaining Voltage 30V
Maximum Current
  I_c Collector Current .... 100 mA

DETECTOR APPLICATIONS

Detectors are extremely versatile in that they may be used with a great variety of electronic devices to perform an equally great variety of tasks. In the figure 4–5, an electroluminescent device is applied to fiber optic light communication. An input square wave signal is driven into a transmitter module. The transmitter module converts the square wave to a current flow for the diode. The diode is an injection laser diode (ILD). When current flows through the diode, the diode radiates a high-intensity light that is launched into the optical cable through the source-to-fiber connector. The light wave is transmitted by reflection to a detector on the receiver side. The detector is probably a photodetector diode such as the PIN diode. The light wave is coupled from the optical cable by a fiber-to-detector connector. The detector converts the light energy to a current flow. The receiver module converts the current to a replica of the original square wave input signal. The LED and the ILD devices are described and illustrated in chapter 6, Optoelectronic Devices and Optical Components.

In the figure 3–6A, a bulk photoresistor is placed on the input leg to an operational amplifier. The light intensity varies the resistance, which varies the input to the amplifier. R_F is a feedback resistor, while R_1 sets the operating level of the operational amplifier. Figure 3–6B, illustrates a PN-junction solar cell in the input leg of an operational amplifier. As in the photoresistor operation in figure 3–6A, the intensity of the light causes an input change to the operational amplifier.

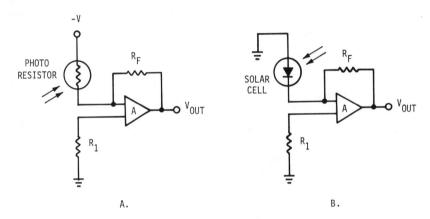

Figure 3–6 Photoresistor and Solar-Cell Detectors (*Mike Anaya*)

In Figure 3–7A, a photodetector transistor is used to drive a relay coil. Light beams applied to the base forward bias the transistor. The transistor turns on, energizing the coil of the relay. Relay contacts close and remain closed until the light beam is removed from the transistor base. In figure 3–7B, a phototransistor is used to drive current through a load. Light beams applied to the base forward bias the transistor. The transistor turns on, causing current to flow through a load resistance. In figure 3–7C, a phototransistor is used as a square wave switch. Light is applied to the base of the transistor in pulses. The transistor turns on and off as the pulses are applied and removed. The output is taken off a resistor in the emitter circuit of the transistor. The output represents the on/off time of the light pulses.

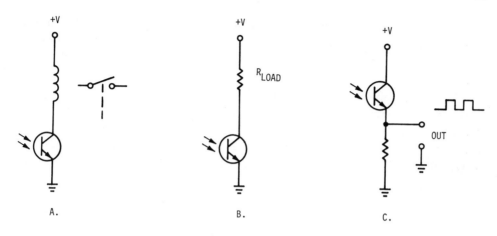

Figure 3–7 Phototransistor Applications (*Mike Anaya*)

In figure 3–8A and B, an electroluminescent device is used to drive a photodetector. The two circuits are similar in that a light-emitting diode (LED) radiates a beam of light that is detected by a phototransistor. In figure 3–8A, the detector is an NPN phototransistor. A second NPN transistor is forward-biased by the phototransistor, which causes current flow through a load in its collector circuit. In figure 3–8B, the detector is an NPN phototransistor. A PNP transistor is forward-biased by the photodetector transistor causing current to flow through a load in its collector circuit.

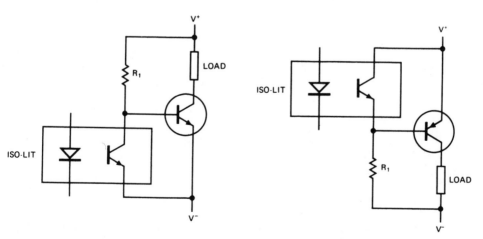

Figure 3–8 Electroluminescent Devices Used to Drive Photodetection Devices (*Courtesy Litronix, Inc.*)

The circuits (devices) illustrated in figure 3–8 are called *opto-isolators.* Opto-isolators are described in detail in chapter 6. The light-emitting diode (LED) is an electroluminescent device. Electroluminescence is also defined in chapter 6.

In figure 3–9A, a phototransistor (part of an opto-isolator) is used to drive TTL (transistor-transistor logic) directly. In figure 3–9B, the NPN phototransistor biases a second NPN transistor which, in turn, drives the logic. More details on the construction and operation of photodetector devices are found in chapter 6.

Figure 3–9   Phototransistors Used to Drive TTL Logic *(Courtesy Litronix, Inc.)*

4

ELECTROLUMINESCENCE

Probably the most useful of all the optoelectronic effects is electroluminescence. This effect involves the application of electrical energy, in the form of current flow, to a semiconductor material causing it to emit light. This light, in the form of photons, is used in applications such as light-emitting diodes (LED's) and in LED combinations to provide digital and alphanumeric displays. Another application of electroluminescence is the injection laser diode (ILD) used to launch light waves in optical fibers.

GENERAL

Electroluminescence is a photoelectric effect in which electrical energy is converted to light energy. This is, of course, the opposite phenomenon to photodetection, where light energy was converted to electrical energy.

Figure 4–1 represents the fundamental operation of an electroluminescent device, the light-emitting diode (LED). Voltage is applied to a circuit by a battery or some other direct-current power source. The circuit contains a load resistance and an electroluminescent device (LED). The load resistance limits the amount of current that can be drawn in the circuit. The LED converts the electron flow to radiant energy. The LED is forward-biased, that is, the battery polarity is applied positive on the P material, negative on the N material. Current is caused to flow through a load resistance into negatively (N) charged material. Holes (absence of electrons) are created by electrons drawn from the positively (P) charged material to the positive pole of the battery. An energy gap is developed between

Figure 4–1 Electroluminescence (*Mike Anaya*)

the negatively charged material and the positively charged material at the junction. It is at this junction that electrons combine with holes to create photons. The electron-hole combinations cause the LED device to radiate energy. This radiated energy is seen through a glass or plastic envelope similar to a small light bulb.

The other electroluminescent device is the injection laser diode (ILD). The ILD is a heterostructure device. Electrons are injected into a junction region where there are deficiencies of free electrons in lattice structures where electron-hole combination takes place. Photons are emitted. Mirror ends reflect the photons back into the active region where more electron-hole combinations take place. The mirror ends serve as a feedback mechanism, making the ILD a semiconductor laser. Control of the brightness is made by adjusting the current flow. The details of the chemical makeup of the LED and the ILD, along with an explanation of the electron-hole combinations, are found in chapter 6, Optoelectronic Devices and Optical Components.

Electroluminescent Materials   The materials that are used for manufacturing solid-state electroluminescent devices are called *compound semiconductors*. This material is chosen by its band-gap energy and its radiation wavelength. There are two common electroluminescent devices, the light-emitting diode (LED), often called a solid-state lamp (SSL), and the injection laser diode (ILD). The basic materials that electroluminescent devices (LEDs) are made of are gallium arsenide (GaAs), gallium phosphide (GaP), and gallium arsenide phosphide (GaAsP).

Gallium arsenide (GaAs) has an energy gap of around 1.4 electron volts at a wavelength of 900 nanometers. Gallium phosphide (GaP) has an energy gap of around 2.2 electron volts at a wavelength of 560 nanometers. Gallium arsenide phosphide (GaAsP) has an energy gap of around 1.9 electron volts at a wavelength of 650 nanometers. These are estimated values and you will find that each manufacturer's LEDs will vary somewhat in wavelength and energy gap.

The first material, gallium arsenide (GaAs), does not emit visible radiation. The spectral wavelength range of GaAs material falls in the infrared range. Although the basic material GaAs emits radiation at 900 nanometers, it may be doped with other materials to provide a wider range of wavelengths. Infrared wavelengths cannot be seen, but are generally included in the classification of LED. Some manufacturers call them infrared LEDs or simply IR LEDs. Others call them infrared emitters.

Gallium phosphide (GaP) emits a green light from 520 to 570 nanometers with its peak around 560 nanometers. It can also radiate a red light between 630 and 790 nanometers with a peak around 690 nanometers. The gallium arsenide phosphide (GaAsP) emits light over an orange-red range depending on the amount of GaP in the material. The GaAsP emits red light between 640 and 700 nanometers with its peak around 650 nanometers. With the correct amount of GaP in the material mix, a yellow light is emitted with a peak around 610-nanometer wavelength.

New compounds of material are coming into operation. For instance, the gallium aluminum arsenide (GaAlAs) has been available for some time. It emits light over a red range from 650 to 700 nanometers with a peak around 670 nanometers.

The efficiency of the LED is very dependent on the emitted wavelength, with drastic fall-off in efficiency as the wavelength gets shorter. The basic material that the injection laser diode (ILD) is constructed from is aluminum gallium arsenide (AlGaAs). The AlGaAs emits infrared light from around 838 to 844 nanometers with a peak near 840 nanometers. The efficiency of an injection laser diode (ILD) lies in its ability to operate in a narrow bandwidth with high intensity.

Direct-Gap versus Indirect-Gap Materials   Electroluminescent devices are used as light radiators because they are direct-gap materials. Other semiconductor materials are indirect-gap materials. Direct and indirect gap refers to the transition of electrons from the conduction band to the valence band. Electron transition takes place when electrons move through a PN junction from N material to P material and combine with holes. When the electrons move from the high-energy state of the conduction band to low-energy state of the valence band, photons of energy are released.

In the direct-gap concept the electrons and holes have equal momentum. Electrons in the conduction band and holes in the valence band may directly recombine. In the indirect-gap concept electrons and holes do not have the same momentum. Electrons and holes are captured or trapped by secondary donor and acceptor levels so that direct transition cannot happen. The electrons and holes do combine and some radiation may occur, followed by transition to the valence band. Direct-gap transition is in one step from the conduction band to the valence band. Indirect-gap transition is in two steps, from the conduction band through an entrapment area to the valence band. The GaAs and the GaAsP diodes are direct-gap radiators. The GaP diode is an indirect-gap radiator.

PERFORMANCE CURVES

Figure 4–2 is a set of typical performance curves for an electroluminescent device. This particular set of curves is for an ITT Model T–820 Stripe Geometry GaAlAs double-heterojunction edge-emitting LED. Other electroluminescent devices have similar curves.

Spectral Response Curve   The first curve in the illustration is the spectral response curve on the top left side. The curve plots wavelength in nanometers against relative intensity. Note that the peak relative intensity occurs at 840 nanometers. The spectral width is around 40 nanometers. The spread is from 825 to 865 nanometers with maximum intensity between 830 and 860 nanometers. The wavelength range of operation is in the infrared. Some of the typical wavelengths versus relative intensity values for other electroluminescent materials are listed below.

Material	Color	Wavelength Peak (nanometers)
GaP	High red	700
GaP	Green	560
GaAs	Infrared	900
GaAs	Infrared	940
GaAsP	Infrared	940
GaAsP	Green	540

Figure 4–2 Performance Curves for Electroluminescent Devices (*Courtesy ITT Electro-Optical Products Division*)

Operating Time/Power Curve The curve in the top right side of the illustration plots operating time in hours against relative output power. The reader will note that the number of hours operating at the limits of the curve is around 7000 hours. The percent of power loss in that amount of time is only several percent.

Forward Current/Output Power Percent Curve The lower left curve in the figure plots forward current in milliamps (mA) against percentage of rated average output power. You will note that this is an extremely linear curve that depicts a direct change in power for increments of current.

Diode Current Voltage Curve   The curve on the lower right side of the illustration is a familiar curve used for all diodes. It plots forward and reverse voltage against current. As voltage increases in the forward-bias direction (right on the curve) current does not flow till around 1.3 volts is reached. In reverse bias current flows slightly until a reverse bias voltage reaches around −1.9 volts. At that point and further, current will increase rapidly in the reverse direction until the diode is destroyed.

Other Performance Curves   There are other performance curves. Case temperature versus normalized output is one and temperature changes due to forward voltage as a function of forward current is another.

SPECIFICATIONS OF ELECTROLUMINESCENT DEVICES

Table 4–1 is a standard set of specifications for an electroluminescent device. This particular set of specifications is for an ITT Model T–820 Striped Geometry GaAlAs double-heterojunction edge-emitting LED. Other electroluminescent devices have similar specifications. Case temperature is a standard as are forward and reverse voltage and continuous and pulsed current. This particular device is used with fiber optics, therefore the optical power output is specified for different optical fiber use. A final useful specification for most radiating devices is the size of the emitting area.

ELECTROLUMINESCENT DEVICE APPLICATIONS

Visual indications are by far the most versatile and useable applications for the electroluminescent device. In figure 4–3A, an LED is utilized in a simple DC circuit. The switch S1 energizes the circuit. Forward bias is applied to the LED as current flows through the load. The LED indicates that current is flowing through the load. In figure 4–3B, the LED is placed in series with a current-limiting resistance and an inverter output. When the inverter output is low (ground) the LED is forward biased and comes on. The current-limiting resistance limits current to the LED.

In figure 4–3C, the LED is placed in series with a current-limiting resistance and in shunt with an inverter output. When the inverter output is high (positive) the LED is forward-biased and comes on. The current limiting resistance

TABLE 4–1 Specifications for ITT Model T-820 Striped Geometry GaAlAs
Double-Heterojunction Edge-Emitting LED.

CHARACTERISTICS
(Case Temperature = 25°C)

MAXIMUM RATINGS

Continuous forward current	100 mA
Pulsed forward current (10 kHz, 50% duty cycle)	200 mA
Reverse voltage	2.0 V
Operating & storage temperature range	−40 to 75°C

	NOMINAL	UNITS	CONDITIONS
Peak emission			
Wavelength	840	nm	$I_F = 100mA$
Spectral width (Full width half max)	40	nm	$I_F = 100mA$
Forward voltage	1.5	V	$I_F = 100mA$
Thermal impedance (Chip to external heat sink)	20	°C/W	
Optical power output			
unencapsulated (T-820)	400	μW	$I_F = 100mA$
coupled into ITT T-102 step-index pigtail (T-821)	55	μW	$I_F = 100mA$
coupled into ITT T-202 graded-index pigtail (T-822)	40	μW	$I_F = 100mA$
coupled into ITT T-302 125 μm core PCS pigtail (T-823)	100	μW	$I_F = 100mA$
Rise/fall times (10% to 90%)	15	ns	
Emitting area	1×20	μm	

limits current to the LED. When the inverter output is low, the LED is shorted
and does not light. In figure 4–3D, the LED is placed in the collector circuit of a
transistor lamp driver circuit with a load resistance. Switch S1 represents some
action such as a door opening or a liquid at a specified level. The switch closes
and forward biases the transistor base-emitter junction. This action causes the
transistor to saturate and in turn the LED comes on.

The electroluminescent device is especially useful in opto-isolator applications (See figure 4–4). In the figure 4–4A, an NPN transistor is used as a driver
for an LED that is part of an opto-isolator. Logic T²L HI input supplies forward
bias to the transistor. The transistor turns on to saturation, supplying a ground
for the LED in the opto-isolator. When current flows through the LED, it radiates
a beam of light. The photo transistor detects the light on its base causing output
current (I_0) to flow.

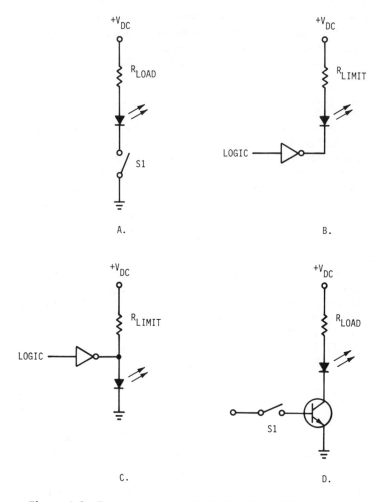

Figure 4–3 Electroluminescent Applications (*Mike Anaya*)

In the figure 4–4B, a PNP transistor is used as a driver for an LED that is part of an opto-isolator. Logic T²L LO input supplies forward bias to the transistor. The transistor turns on to saturation, supplying a positive voltage for the LED in the opto-isolator. When current flows through the LED, it radiates a beam of light. The photo transistor detects the light on its base causing output current (I_o) to flow.

In the figure 4–4C, a buffer NAND gate is used as a driver for an LED within an opto-isolator. When the NAND gate has a low output, a ground is supplied for the LED. Current flows through the LED. The LED radiates a beam of light. The phototransistor detects the light on its base causing output current to flow.

Figure 4-4 Electroluminescent Devices Used as Opto-Isolators (*Courtesy Litronix, Inc.*)

In the figure 4–5 an electroluminescent device is applied to fiber-optic light communication. An input square wave signal is driven into a transmitter module. The transmitter module converts the square wave to a current flow for the diode. The diode is an injection laser diode (ILD). When current flows through the diode, the diode radiates a high-intensity light that is launched into the optical cable through the source-to-fiber connector. The light wave is transmitted by reflection to a detector on the receiver side. The detector is probably a photodetector diode. The light wave is coupled from the optical cable by a fiber-to-detector connector. The detector converts the light energy to a current flow. The receiver module converts the current to a replica of the original square wave input signal. The LED and the ILD devices are described and illustrated in chapter 6, Optoelectronic Devices and Optical Components.

Figure 4–5 Electroluminescent Device Used with Fiber Optics (*Courtesy ITT Cannon Electric*)

5

SAFETY

GENERAL

Safety education today has become an important phase of everyone's life. Under the 1970 federal Occupational Safety and Health Act (OSHA), the employer is required to furnish a place of employment free of known hazards likely to cause death or injury. The employer has the specific duty of complying with safety and health standards as set forth under the 1970 act. At the same time, employees also have the duty to comply with these standards. A full treatment of the subject of safety is far beyond the scope of this book, and there is ample justification for a full course in safety procedures, including first-aid treatment, for all electrical and electronic technicians. The intent of this chapter is to make the reader fully aware of the ever-present, invisible, and generally silent hazards in handling electrical apparatus and to point out some fairly common causes of electrical shocks and fires that can easily be overlooked.

Since the time when Benjamin Franklin flew his famous kite, it has become more and more apparent that electricity, even in its milder forms, is dangerous. A fact not widely mentioned in history is that shortly after Franklin's experiment a Russian experimenter was killed in an attempt to duplicate the kite trick. We can therefore assume that good safety habits are mandatory for all who use, direct the use of, or come in contact with electricity. Electrical equipment is found in every place that the ordinary person may find himself or herself. For this reason it becomes the responsibility of each of us to be knowledgeable of electrical safety and to become our own and our brothers' keepers.

The most basic cause of electrical accidents, as other types of accidents, is carelessness, and the best prevention is common sense. However, knowing exactly what to do in an emergency is only achieved through formal education or experience. Unfortunately, the experience could be fatal, so it is more desirable to derive your knowledge through schooling. When an emergency happens it is often accompanied by panic that can cause the mind or muscles to become paralyzed. The antidote for this panic is to be safety conscious.

SAFETY RESPONSIBILITY

The supervisor is responsible for enforcing the rules of safety in the area under his or her direct supervision. Inspectors ensure that equipment is tested before it is released for use, and final testing should include safety precautions. Finally, the user of the equipment should be qualified in its operation so he or she knows when it is operating in the proper manner. No one who comes into contact with such equipment is exempt from some degree of responsibility. The average layperson may not believe he or she can get a severe shock from an electrical component disconnected from a power source, but even that could present a hidden danger.

ELECTRICAL SHOCK

There is a common belief that it takes a great volume of electricity to cause a fatal shock and that high voltage is the thing that provides the jolt to do the job. Imaginative stories about people being literally fried or jolted from their shoes are supplied freely by storytellers. Although there are true tales of this type, this is not the total picture. The bulk of electrical shocks come in small packages. Death from electrical shock happens most often from ordinary 60-hertz, 115-volt alternating-current house power. The effect, in general, is an instantaneous, violent type of paralysis. The human body contains a great deal of water and is normally somewhat acid and saline. For these reasons it is a fairly good conductor of electric current, which has no regard for human life or feelings.

Our brains are loaded with tiny nerve ends that provide a transmitting service to the muscles. Muscles, in turn, provide motion for the body functions such as the heart and lungs. Now, assume that an electrical impulse that was not called for told your heart to speed up or stop its pumping job. Or suppose that a similar impulse of electricity told the lungs to quit taking in air. These situations do occur and with a comparatively small amount of current flow. Depending on a multitude of complex variables, a small current such as 10 milliamps can be very unpleasant. Currents no larger than 20 milliamps can cause muscle tightening or freezing. Currents of 30 milliamps can cause damage to brain tissue and blood vessels. And damage to brain and blood vessels can, of course, be fatal.

As you know, a decrease in resistance (according to Ohm's law) causes current to increase. Body resistance decreases when perspiring or otherwise wet. A great number of things can cause variation in resistance. The general health of the person is probably the most important variable. A body in good condition has a much better chance of recovering from electrical shock. The muscles, by being in better tone, can recover to normal from paralyzation.

The path of current flow varies the shock. For instance, if the current involves the brain or heart it is naturally more dangerous than another path. The length of exposure can also be a factor, as well as the size or surface area of the electrical contact. Large voltages can cause spastic action, but recovery can be rapid. Also, small currents can cause muscle paralysis. In the event of breathing paralysis, artificial respiration must take place as quickly as possible to prevent loss of body functions and damage to the brain because of lack of oxygen. A condition known as ventricular fibrilation (uncoordinated heart beats, both fast and irregular) can occur with high currents, say 50 to 100 milliamps. This action will continue until something is done externally to restore regular heartbeat, that is defibrilation or CPR (heart massage).

Death normally occurs when currents reach 250 milliamps. This does not have to occur, however, as rapid first aid can save a victim whose heart has actually stopped beating or whose lungs have stopped pumping momentarily. In many cases rapid action by knowledgeable persons can prevent body damage and save lives.

RAPID RESCUE TECHNIQUES

A special course in first aid will ensure the proper methods. A wrong method can be worse than no action at all. The first rule of thumb for an electrical shock victim is to remove the current path. This can be done by turning the power switch off if it is readily accessible. If not accessible, the person can be detached from the current source by using an insulator of some sort such as wood, rubber, cork, or plastic. Sometimes it is more suitable and sensible to remove the power source from the victim. Whichever is the case, isolation from the current source is by far the most important and first move that can be made.

Isolation procedures can cause problems. The rescuer may find himself or herself in the current path. Touching a person who is paralyzed to a current source provides the current source another path through the rescuer's body. Care must be taken to prevent this from happening. After isolation, artificial respiration and/or other first aid should be applied. In all cases speed is vital. The likelihood of an electrical shock being fatal increases the longer the victim remains without breathing. A person given artificial respiration in the first 3 minutes has a much better chance of survival than one who is given artificial respiration after 5 minutes.

SAFETY IN OPTOELECTRONICS

The biological effects of electromagnetic radiation to human beings fall into two categories. These are ionizing and nonionizing radiation. The first, ionizing radiation, is related to cosmic and x-ray wavelengths and to nuclear radiation (very small). The second, nonionizing radiation, is related to ultraviolet, visible, infrared, microwave, and radio wavelengths (very large). The larger a wavelength is, the smaller its frequency, and vice-versa. To distinguish the exact position in the frequency spectrum where the two radiation categories (ionizing and nonionizing) fall is impossible. Each frequency and wavelength level has an overlap to neighboring areas. The dividing point between optical and microwave is usually established around 1 millimeter wavelength. The visible spectrum falls in the optical area in the range of 380 to 760 millimeters. The exact level is subject to disagreement and depends a great deal on whose book you are reading. In any event, ionizing radiation by scientific analysis is the minimum photon energy required to produce ionization in the atoms of biological tissue or around 103 nanometers (ultraviolet). The word *ionize* means to change atoms to ions. Nonionizing radiation is, of course, not involved with this process. The optical levels of the frequency spectrum are of the most significance in this study of optoelectronics. The writer does not wish to iterate that health hazards do not exist in other levels of radiation. If exposed to enough radiant energy of almost any type damage to the human body may be bothersome, painful, critical, and perhaps final.
      Since we are mainly concerned with the optical spectrum in this book, we shall limit the radiation safety discussion to nonionizing radiation.

BIOLOGICAL EFFECTS OF RADIATION

Modern developments in optical technology include fiber optics, the laser, and peripheral light sources such as electronic flashlamps and arc lamps have caused safety engineers some concern. The Division of Industrial Safety and OSHA have changed their standards to include these high-energy light devices. Beside the damage possible to the eyes, many of the strong light sources are capable of producing hazardous levels of ultraviolet radiation. OSHA standards are now set to include radiation that produces injury to the skin and the eyes. Prior to 1960, public use of microwave radiation was extremely limited. Changes in high-power radar equipment, microwave communications, industrial heating, and home microwave cooking have exposed the public to radiation that they previously had not encountered. The expansion of the fiber optics industry and the increasing use of new laser devices since the beginning of the 1970s has brought forth empirical data that have caused changes in safety regulations.

Biological effects of nonionizing radiation are dependent on the spectral region of the radiation and the duration of the exposure to the radiation. Furthermore, the damage to the eyes and skin is dependent on whether there was just a single exposure (acute) or daily exposure (chronic) to the radiation. Any discussion on permissible radiation exposure should include these factors. Quantities are usually expressed in exposure doses and exposure dose rate. Physical factors also affect the radiation hazard. These are, the radiation absorbed in, reflected from, and transmitted through the tissue. A final physical factor that must be considered is the spectral sensitivity of the particular tissue exposed to the radiation.

The biological effects of optical radiation to the human body are thermal (heat) and nonthermal. The eye is probably the organ that is most sensitive to injury. The injury to the eye is dependent upon the region of the eye that absorbs the radiant energy.

Radiation and the Eye   The eyes, as we have stated, are generally considered to be the organ of the body which is most susceptible to damage by radiation. The parts of the eye that can be affected by radiation are the cornea, lens, eye fluid, and retina (see figure 5–1). Different light radiation affects the individual eye parts. The damage to any of the parts occurs when the light is absorbed by the parts. The damage that takes place is dependent on the ranges of the exposure levels and the time of exposure.

Figure 5–1 The Eye and Its Parts (*Mike Anaya*)

In the optical spectrum, wavelengths for ultraviolet radiated frequencies range from 0.005 microns to 0.3900 microns. Short wavelengths of ultraviolet radiation are absorbed by the cornea. Long wavelengths of ultraviolet radiation are generally refracted at the cornea and the lens then absorbed in the retina. Some long wavelengths of ultraviolet radiation are absorbed by the cornea and lens.

Visible wavelengths of radiated frequency range from 0.3900 microns to 0.7500 microns. Visible wavelengths are generally refracted by the cornea and absorbed by the retina. Infrared wavelengths of radiated frequencies range from 0.7500 microns to 4000 microns. Near-infrared wavelengths are absorbed primarily in the ocular medium and at the retina and for that matter all parts of the eye. The near-infrared frequencies are the most dangerous to the eye. Far-infrared frequencies are absorbed at the cornea.

ANSI Z126.1 Standards for the Safe Use of Lasers   The ANSI (American National Standards Institute) standard has divided lasers into classes which represent their possible hazards.

Class I lasers are at visible wavelengths. These are considered to be safe for continuous viewing at irradiance levels of 0.4 microwatts passing through the cornea.

Class II lasers are also at visible wavelengths. They are considered as those lasers that do not have enough power to cause retina damage during momentary viewing. Prolonged viewing will cause damage no matter what the irradiance levels are.

Class IIIA lasers will not generally cause damage to the eye under normal conditions, but will damage the eye when viewing the beam through optics. These are medium-power lasers in the range of 1.0 to 5.0 milliwatts.

Class IIIB are lasers which cannot cause hazardous diffuse reflection or fire. This medium-power class ranges from 5.0 to 500 milliwatts; they can cause direct eye damage.

Class IV lasers are a diffuse reflection hazard and a fire hazard. They are classified as high power and range from 500 milliwatts upward.

Special tags with the class and its peculiarity are required to be posted. Samples of these tags are illustrated in figure 5–2.

Lasers in general must be handled with extreme caution. General safety rules include prevention of direct sight at the beam, the avoidance of reflective

Figure 5–2 Laser Safety Tag Samples (*Courtesy Coherent, Inc.*)

objects such as mirrors, jewelery, etc., and the access control to keep unqualified persons from harm's way. Lasers should only be used in well-defined areas. Appropriate signs must be posted in applicable places to alert people of the potential hazards. Whenever the laser is not in use, the laser room should not be accessible to persons who are not educated in laser technology. Accidents should be immediately reported. Even in minor eye accidents, an ophthalmologist should be consulted.

Problems of High-Energy Directed Light In fiber optics and laser technology, the problems of radiation are compounded by extremes of density (high energy) and direction. Light beams in fiber optics are transmitted in glass fibers of 4 or 5 micrometers in diameter. Beam sizes of lasers are common in the 8- to 10-micrometer range. The hazards exist when an extremely strong beam of light is focused or is looked at directly. The beam, with its massive irradiant strength (often called power density) passes through the pupil and is directed in force upon the retina. Irreparable damage may be done in an extremely short period of time. If the beam is strong enough, damage can occur when the light beam is seen indirectly such as when bounced off or reflected from a smooth surface.

FEDERAL REGULATIONS

In Title 8, the Division of Industrial Safety provides general safety orders that are to be adhered to by industry.

Eye and Face Protection Employees working in locations where eye hazards due to flying particles, hazardous substances, or injurious light rays are inherent in the work or environment, shall be safeguarded by means of face or eye protection. Suitable screens or shields isolating the hazardous exposure may be considered adequate safeguarding for nearby employees.

The employer shall provide and the employee shall use protection suitable for the exposure.

Where eye protection is required and the employee requires vision correction, such eye protection shall be provided as follows:

1. Safety spectacles with suitable corrected lenses, or
2. Safety goggles designed to fit over spectacles, or
3. Protective goggles with corrective lenses mounted behind the protective lenses.

The wearing of contact lenses is prohibited in working environments having harmful exposure to materials or light flashes, except when special precautionary procedures, which are medically approved, have been established for the protection of the exposed employee.

Design, construction, testing, and use of devices for eye and face protection shall be in accordance with American National Standard for Occupational and Education Eye and Face Protection, Z87.1–1968, except that integral lens and frame design will be allowed if the lens-frame combination provides unit strength as well as impact penetration, heat and flammability resistance, optical qualities, and eye zone coverage equal to or greater than is required by ANSI Z–87.1 1968.

Employees whose occupation or assignment requires exposure to laser beams shall be furnished suitable laser safety goggles which will protect for the specific wavelength of the laser and be of optical density (O.D.) adequate for the energy involved.

The ANSI standard Z136.1 divides lasers into classes dependent on the radiation hazard.

6

OPTOELECTRONIC DEVICES AND OPTICAL COMPONENTS

GENERAL

Optoelectronic devices are, of course, those devices that respond to radiation of light or emit radiation. They either convert electrical current to radiant energy or convert radiant energy to electrical current. Devices that convert electrical current to radiant energy are called *electroluminescent devices*. There are three device categories that convert radiant energy to electrical current. Photoemissive devices emit electrons when radiant energy strikes sensitive material. Photoconductive devices change resistance due to radiant energy. Photovoltaic devices generate a voltage due to radiant energy. The term radiation includes ultraviolet, visible, and infrared light in the optical spectrum. Photometric terms are those that deal with the visible wavelengths. Radiometric terms are those that deal with wavelengths outside the visible spectrum. This includes wavelengths from 0.005 microns to 4000 microns. In frequencies, these are high values (6×10^{16} hertz to 7.5×10^{10} hertz). This chapter of the book will deal with various light sources with emphasis placed on the solid-state devices. The devices covered may operate at ultraviolet, visible, or infrared wavelengths. Along with the individual devices, we shall introduce some of the displays including the intelligent display. This is the state of the art in optoelectronic devices. We shall also cover opto-isolators, another valuable device for conversion to and from light energy.

We shall finally take a look at some of the more important optics such as windows, mirrors, reflectors, beamsplitters, and polarizers. It must be understood that the chapter provides only a cursory look at devices and optics. The ones in this chapter are major to optoelectronics and specifically to fiber optics and laser technology.

PHOTODIODES

One of the most basic of all the optoelectronic solid-state devices is the photodiode. The photodiode may be used as a photovoltaic device or a photoconductive device. A photovoltaic device, you may recall, converts radiant energy to a generated voltage. In a photoconductive device, electrical resistance is changed when exposed to radiant energy. You may wish to refer to chapter 3 for a review of photovoltaic and photoconductive action. There are two photodiode types on the market today. These are the conventional PN junction diode and the PIN diode.

Conventional PN Junction Photodiode Operation   Figure 6–1 is a cross section drawing of a conventional photodiode. There have been many processes by which PN junctions have been formed. The technique that has risen to be the most popular and the one utilized most by industry is the planar process. This process is utilized by integrated-circuit manufacturers. The process uses silicon as the solid and gases as the dopants. By diffusion, the dopant gases penetrate the solid surface of the silicon. Diffusions may be made on diffusions, therefore several layers of dopants may be diffused on one device. This serves to provide manufacturing versatility.

Figure 6–1 Cross Section, Planar Diffused Photodiode (*Courtesy United Detector Technology, Inc.*)

The basic PN junction used in production of the photodiode is the planar diffused. In the illustration, N-type bulk silicon is diffused on one side by N+ dopant and on the opposite side by P+ dopant. A depletion region between the N and P exists free of current carriers. It is in this depletion area that the photons should be absorbed. To operate in the photoconductive mode the device must be reverse biased. To operate photovoltaically, no bias is required. An active area exposes the P+ diffusion to light beams. The light beams are absorbed into the semiconductor. When photons of energy are absorbed into the material, electron-hole pairs are formed. Short, medium, and long wavelengths of photon power are absorbed at different depths within the PN junction. The depth is dependent on the photon wavelength. Short wavelengths, of course, are absorbed near the surface. Long wavelengths may penetrate the entire structure. To be most useful, the wavelengths should be absorbed in the depletion area.

Current is produced by electron-hole pairs being separated and drawn out in directions of more positive or negative sources whichever is the case. If the electron-hole pairs happen outside the depletion area they will usually combine and no current will be produced. The active region (P+ diffusion) should be extremely thin to ensure maximum penetration. As in other reverse-biased diodes, the depletion area can be made larger by increasing the reverse bias.

In the figure 6–2, the Schottky barrier photodiode is illustrated. Operation of the Schottky photodiode is much the same as the conventional PN junction. The Schottky diode (often called a surface diode) differs in the method by which the P-type material is formed. In the conventional photodiode, the method is diffusion. In the Schottky photodiode, the method is by metallization. In the case illustrated, the active region has a thin gold film metallized to the N-type silicon bulk. The Schottky barrier photodiode has some advantages over the conventional photodiode. It operates well at wavelengths less than 500 nanometers and has much simpler fabrication processes. The Schottky photodiode does not operate well at high temperatures or with high light power.

Figure 6–2 Cross Section, Schottky Barrier Photodiode (*Courtesy United Detector Technology, Inc.*)

The PIN Photodiode   The PIN diode is so called because of the layer material by which it is constructed. The word PIN is an acronym for P-type, Intrinsic, N-type materials. A PIN photodiode is one in which a heavily doped P region and a heavily doped N region are separated by a lightly doped I region. In figure 6–1, the N-type silicon bulk would represent the I region. The resistance of the I region can range from 10 ohms per centimeter to 100,000 ohms per centimeter. The P and N regions are less than 1 ohm per centimeter. Since a depletion area can extend further into a nondoped or lightly doped region the PIN photodiode has an extremely large depletion area. This large depletion area provides the PIN photodiode with much faster speeds, lower noise, and greater efficiency at longer wavelengths.

Spectral Response Curves   Each manufacturer issues spectral response curves which provide the range of wavelengths in which their devices operate (see figure 6–3). These curves usually have a shape that corresponds to the CIE photopic response curve. Response is based on the photodiode's ability to absorb photons of light energy. Longer-wavelength photons are so low in energy that they can penetrate the entire photodiode structure without producing an electron pair. Therefore there is a response cutoff as illustrated on the curves. Short wavelengths in the ultraviolet area fall off because electron-hole pairs are combined in the active area of the photodiode without reaching the depletion area. Visible wavelengths have higher response with Schottky barrier than planar diffused photodiodes.

Photodiode Applications   Photodiodes are used in photoconductive and photovoltaic applications. The characteristics illustrated in figure 6–4 are for a planar diffused photodiode. In the photoconductive mode the device must be operated with reverse bias. The problem with reverse bias is that the device develops reverse current, called dark current. The dark current is present with reverse bias when no outside radiation is applied.

In the photovoltaic mode the device operates with zero bias. This makes sense since the photodiode in this mode is the generator producing the load voltage. On the top left side of the illustration, the important spectral response curve is provided. You will note that this curve is similar to the one previously illustrated. This particular curve is for planar diffused silicon photodiode. The curve has the typical response cutoff at long wavelengths. Short-wavelength cutoff is at ultraviolet wavelength.

A second curve that is meaningful to a photodiode operation is the linearity curve. This curve compares incident radiant energy with photocurrent in amperes.

Figure 6-3 Spectral Response Curves (*Courtesy United Detector Technology, Inc.*)

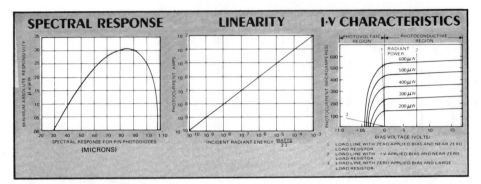

EQUIVALENT CIRCUIT (all models)

TYPICAL HOOKUPS

Figure 6–4 Planar Diffused PIN Diode Characteristics (*Courtesy United Detector Technology, Inc.*)

Note that the curve is extremely linear. When the photodiode is exposed to radiant energy, valence electrons are released. The greater the radiant energy exposure the greater the current flow.

The current-voltage (I–V) characteristic curve is representative of both photovoltaic and photoconductive operation. Note that photovoltaic region of bias voltage is at zero bias with the photodiode generating the load voltage. This is called the *voltage mode*. The photoconductive region requires constant reverse bias. This is called the *current mode*. Current is extracted as a measure of applied radiant energy.

The equivalent circuit in the center of the illustration is representative of all photodiodes. The dark current (I_o) is caused by reverse bias. The signal current (i_s) is the current developed by input light. Noise current (i_n) varies with bandwidth, temperature, and dark current. An important ratio for designer use is the signal-to-noise ratio (i_s/i_n.) It is obvious that the greater the signal-to-noise ratio the better the design.

Typical hookups are illustrated on the lower part of the illustration. The photoconductive circuit has reverse bias, photovoltaic does not. Signal current flows through the load resistor R_L in the photoconductive circuit. An operational amplifier is used in the photovoltaic circuit to amplify the signal from the photodiode. The input resistance of the operational amplifier is extremely high. The resistance (R_f) is used for feedback.

A planar diffused photodiode is illustrated in figure 6–5. This particular photodiode is manufactured to operate in the photoconductive mode. That is, the diode must be reverse-biased. The only problem with reverse biasing is that the diode is subjected to the flow of dark current. Dark current you may recall flows when there is no radiant flux on the photodiode. Dark current is an integral part of noise current and must be used in determining signal-to-noise ratio. The dark current for this particular device is extremely small (0.002 microamperes typically with 10-volt bias).

Figure 6–5 Planar Diffused Photodiode (UDT PIN Spot) (*Courtesy United Detector Technology, Inc.*)

The United Detector Technology (UDT) spot diodes are bi-cell or quadrant-cell detectors ideally used for a wide range of nulling and centering applications. The devices consist of two or four discrete elements on a single silicon substrate with an active lead from each element. When a light beam is centered on the detector (null or center position is the intersection of active elements), output current from each quadrant is equal. As the light beam moves, current indicates off-center position. The spot diode exhibits stability over time and temperature, high responsivity, and fast response times per pulse operation. Note that the one photodiode has two discrete elements and the other has four.

The spot diode has broad frequency response, fast rise time, and low barrier capacitance. If these are features that are required for the design, the spot diode must be used in the photoconductive mode regardless of dark current. If dark current is a problem for a design, then the photovoltaic mode should be utilized. The UDT spot diode is applied in lens manufacturing, feedback control systems, guidance systems, laser alignment, machine tool alignment, targeting, and process machinery alignment. It is primarily used in the photoconductive mode of operation and position-sensing detection.

Photodiodes that are manufactured to operate in the photovoltaic mode do so with zero bias. This is ideal for the elimination of dark current in applications that demand this requirement. Photodiodes that are manufactured for photovoltaic modes are used in radiometers, photometers, spectrometers, densitometers, colorimeters, particle counters, fluorescent analysis, replacement of phototransistors, point-of-scale scanners, intrusion alarms, opto-couplers, E–O blood analyzers, and video-disc recorders. Large-area photovoltaic photodiodes have application in solar cells, food processing, illumination control, proximity detection, and automatic conveying.

The Schottky barrier photodiode represents the state-of-the-art in large-area, high-sensitivity, fast-response, expanded-spectral-range detectors. The cold-formed Schottky barrier preserves the high resistivity of the intrinsic (I) region of the PIN diode. This serves to provide low barrier capacitance, low dark current, and low noise. The thin gold contact used in metallization allows ultraviolet and blue light to pass easily into the depletion area where electron pairs are more effectively collected. Schottky barrier photodiodes are used in radiometry, photometry, densitometers, colorimeters, laser power meters, laser range finders, ultraviolet flame detection, character recognition, alignment, and other position-sensing detectors. A variety of photodiodes are pictured in figure 6–6.

Solar Cells   The solar cell is probably the greatest user of the photodiode in the photovoltaic mode. The solar cell converts the sunlight energy to electrical

Figure 6–6 Other Photodiodes (*Courtesy United Detector Technology, Inc.*)

energy. In space the solar cell is used to provide power for spacecraft. The efficiency of the cell at high altitudes is very high. Beyond the earth's atmosphere the sun's rays are not inhibited. Solar cells are being used on the earth for many applications, including remote power supplies.

Solar energy has its peak at around 500 nanometers. Power from this source is nearly 100 milliwatts per square centimeter if it can all be collected. The solar-cell structure is very simple and similar to the photodiode. The depletion area is very narrow. The active area is very large to intercept the maximum radiant flux. Speed of the solar cell is very low. The diode is operated in the photovoltaic mode without bias. Photons of adequate energy create electron-hole pairs. The pairs are drawn from the narrow depletion area with a high probability of recombination. The current created from the recombination is directed through a very low resistance to obtain maximum power transfer.

Solar cells in space may reach 75 percent efficiency, whereas those on the earth are still struggling to obtain 20 percent. There have been indications of a breakthrough in efficiency. Even a jump of 5 percent would be significant. Solar cells are especially useful in remote areas for recharging batteries to operate communication equipment. Solar cells help in operating weather devices.

Cell banks in remote areas must be set to obtain the most sunlight throughout the year. If servos are used to place the solar-cell banks in the right position, the current created by the cell is used up in moving the bank. Therefore, the mechanics of positioning from season to season, along with varying temperatures, cloudy days, and sunlight time, are certainly dominant obstacles against efficiency.

LIGHT-EMITTING DIODE (LED)

The LED represents the best of the electroluminescent devices. Electroluminescence, you may recall, is the emission of light from a solid-state device by application of current flow through the device. The light comes from photons of energy caused by hole-electron combinations. The diode is forward-biased from an external source. Electrons are injected into the N-type solid-state material. Holes are injected into the P-type solid-state material. The injected electrons and holes recombine with majority carriers near the PN junction. The result is radiation in the form of photons in all directions and specifically from the top surface.

Cross Section of an LED   In figure 6–7, a standard red LED is illustrated. In figure 6–8, a bright orange LED is shown. You will note that in general these both have similar cross sections. Both have substrates on which epitaxial layers are grown. Each has a gold (Au) ball bond for current application.

Let's consider some of the materials from which the light-emitting diodes are made. Most are made from so-called compound semiconductors such as gallium arsenide phosphide (GaAsP), gallium phosphide (GaP), and gallium aluminum arsenide (GaAlAs). Gallium arsenide (GaAs) is usually included in this group but it must be understood that GaAs emits only infrared radiation around 900-nanometer wavelength. These cannot be seen, of course, and are not generally classified as LEDs. Some manufacturers call them IR LED's, while others name them infrared emitters.

The GaP emits green light from 520 to 570 nanometer wavelength with its peak around 550 nanometers. It can also emit red light between 630 and 790 nanometers with a peak of 690 nanometers. The GaAsP emits light over an orange-red range depending on the amount of GaP in the material. The GaAsP emits red light between 640 and 700 nanometers with a peak at 660 nanometers. With the correct amount of GaP in the material a yellow light is emitted with a peak around 610 nanometers wavelength. The GaAlAs emits light over a red range from 650 to 700 nanometers with a peak around 670 nanometers. The efficiency of the LED is very dependent on the emitted wavelength with drastic fall off in efficiency as the wavelength gets shorter.

In the cross sections of figures 6–7, and 6–8, you will note that the layers of GaAsP are grown on substrates of GaAs and GaP, respectively. The substrates are grown in ingots as in other solid-state devices. A lattice structure which is pure and free from imperfections is produced under a controlled environment. The epitaxial layers are gradually changed to maintain a perfect crystalline structure. Note that the red LED epitaxial layers are doped with tellurium (Te), while the orange LED epitaxial layers are doped with sulphur (S). The wafer is then

Standard Red GaAs$_{.6}$P$_{.4}$ cross section

Figure 6–7 Red LED Cross Section (*Courtesy Litronix, Inc.*)

GaAs$_{1-x}$ P$_x$ on GaP cross section (Bright Orange)

Figure 6–8 Orange LED Cross Section (*Courtesy Litronix, Inc.*)

coated with a layer of silicon nitride (SiN) which acts as a barrier to the diffused P-layer. The P-layer diffusant is made of zinc (Zn). It is placed in the opening in the silicon nitride (SiN) to form the PN junction. The outer casing on the red LED in figure 6–7 is made of aluminum (Al) on the front side with gold-tin (AuSn) on the backside contact. A gold ball bond is used to establish bias contact.

The outer casing on the orange LED in figure 6–8 is made of gold-beryllium (AuBe) with a backside reflective contact of gold-germanium (AuGe). A gold ball bond is used to establish bias contact. Figure 6–9, is a plan view of a typical standard red LED die (chip).

It must be stated that the manufacturing processes involved in the growing of substrate ingots and the further growth of epitaxial layers, impurity diffusions and thin film deposition is a complex and highly technical task. Since this is the case, most manufacturers have secret or proprietary procedures because of the high cost of research and development involved.

Figure 6–9 Plan View of Typical LED Die (Chips) (*Courtesy Litronix, Inc.*)

LED Lamps   In figure 6–10 are a series of typical LED lamps. LED lamps are made so as to be as bright as possible within their chemical capabilities. The LED chip is placed within a plastic structure that provides a maximum full-flooded front-radiating area and wide-angle viewing. They are also made to be easily soldered on a PC board or snapped into a mounting clip.

Figure 6–10   LED Lamps (*Courtesy Litronix, Inc.*)

The choice of an LED for a specific use is obvious in many ways. The choice usually boils down to five basic factors.

1. The color of an LED is selected by the user for a particular use. Generally the color is in various shades of red and often orange. Amber and green are also available but the difficulty in locating suitable raw material may preclude the choice.
2. The size of the LED depends, of course, on where the device is to be installed. Many combinations of chip and casing sizes are available which range from 5 mils to 300 mils in diameter. The larger the size, the more visible the device radiance.

3. Most LED chips are Lambertian in angular distribution of light. That is, the luminance of the LED can be seen equally well from all directions. With suitable design the angular light pattern may be changed from a very broad pattern to a very narrow pattern. By placing the chip at the focus of the lens system, a narrow high-intensity beam is obtained. By using diffusing lens material a large area source is obtained but luminance is reduced.

4. The luminous intensity of the device will govern the visibility under background contrast conditions when viewed at normal distances. A typical LED lamp has a luminous intensity of 1 millicandela. Luminous intensity is flux per solid angle.

5. When it is not possible to provide a dark contrasting background, or when the source (LED) is viewed at very close distances, the luminance becomes important. Luminance is the intensity per unit area of a source. Values are from 100 foot-lamberts to 5000 foot-lamberts.

The electronics engineer may find the use of photometry and radiometry a confused mass of strange units and values. Chapter 10 of this book strives to organize these units.

LED Characteristic Curves   The characteristic curves in the figure 6–11 are representative of a typical red LED lamp. These particular curves are for high bright gallium arsenide phosphide (GaAsP). The curves represent the following characteristics:

1. *Relative luminous intensity vs. angle*—Relative intensity in % versus the number of degrees off angle in %.
2. *Luminous intensity vs. ambient temperature*—Percent of relative luminance versus the ambient temperature in centigrade (°C).
3. *Spectral distribution*—Relative intensity versus wavelength λ in nanometers.
4. *Forward current vs. forward voltage*—Forward current in milliamperes versus forward voltage in volts of direct current.
5. *Luminous intensity vs. forward current*—Luminous intensity in millicandela versus forward current in milliamperes.

These are typical of LED characteristics and are self-explanatory.

A Flashing Light-Emitting Diode   Litronix FRL 4403 is a gallium arsenide phosphide (GaAsP) solid-state lamp with a red diffused plastic lens. A

**RELATIVE LUMINOUS
INTENSITY VS ANGLE**

**LUMINOUS INTENSITY VS
AMBIENT TEMPERATURE**

**SPECTRAL
DISTRIBUTION**

**FORWARD CURRENT VS
FORWARD VOLTAGE**

**LUMINOUS INTENSITY VS
FORWARD CURRENT**

Figure 6–11 Typical LED Characteristic Curves (*Courtesy Litronix, Inc.*)

built in IC chip flashes the lamp on/off to attract attention. Its pulse rate is 2.5 hertz. The lamp is a T1¾ size and has a wide viewing angle. This lamp is significant because it can be driven by standard TTL and CMOS circuits without the need for external circuitry.

Specifications for the flashing LED are, in general, much the same as standard LEDs. Some of the factors involved are luminous intensity, emission peak wavelengths, spectral half-width, operating voltage and current, and temperature ranges. Pulse rate is, of course, an additional factor.

Applications for the flashing LED are shown in figure 6–12. The first application is directly driven by TTL or MOS buffer gate outputs. The second application is a flashing LED set in parallel with a 5.6-volt zener diode and in series with a current-limiting resistance. This application is for operation at a voltage level of greater than 5 volts. The third application represents two LEDs flashing together. The final application illustrates one flashing LED in the bias leg of a transistor while the second is in the load leg of the transistor. This allows the LEDs to flash alternately.

Figure 6–12 Applications for Flashing LEDs (*Courtesy Litronix, Inc.*)

The Infrared LED A typical infrared LED is made from gallium arsenide (GaAs). This is an excellent material to make the diode from because most of the material's energy is given up in the form of radiant energy. With forward bias the diode emits a narrow wavelength of around 900 nm. Typical specifications include thermal ratings, radiant intensity curves, and forward current and voltage ratings.

Operating LEDs on ac Power Often it is desirable to place an LED in an alternating current (AC) circuit (See figure 6–13). There are basically three aplications typical of the LED operation in ac circuits. In the (A) application an LED is placed in series with a rectifier diode and a series current-limiting resistance. High power is dissipated in the resistor.

In the second application (B), a rectifier diode is placed in shunt with an LED. A capacitor and current-limiting resistor are in series with the parallel diodes. For current in either direction the voltage drop across the LED or rectifier is a negligible part of the applied ac volts. Capacitor current is almost equal to the ac supply voltage divided by the capacitor reactance ($I_{CAP} = V_{AC}/X_C$). The resistor in series is used to limit turn-on transient currents. Power dissipation is very low and is limited to the LED and rectifier losses, resistance losses, and capacitive reactance losses.

In the third application (C), three LEDs are operated from the same source. The LEDs are simply placed in series in the circuit with a shunt rectifier diode. Current is essentially the same as for one LED.

Figure 6–13 LED Applications Using AC Power *(Courtesy Litronix, Inc.)*

Mounting LEDs   The LED comes in standard sizes. The most prevalent sizes used in electronics are the T1 and the T1¾ sizes. These numbers relate to the diameter of the lamp in eighths of an inch. There are, of course, other shapes and sizes, such as the subminiature and the rectangular. Most lamps have plastic lenses, but others may be made with hermetically sealed glass. Some have spherical lenses. As you may realize, mounting techniques are dependent on the shape, size, and use of the LED.

There are many ways to mount an LED. Some of these are by soldering directly into PC boards, plugging in preassembled sockets, panel mounting, or mounting with clips. Lead bending is allowed, but bending must follow manufacturers recommended bend radii. The leads should be clamped next to the base during bending to relieve stresses. Mechanical side loads or other force should not be placed on the LED during installation. The figure 6–14 illustrates functional mounting of LEDs.

Figure 6–14   LED Mounting (*Courtesy Litronix, Inc.*)

Most LEDs are subject to thermal stress. A typical LED chip is mounted on a substrate or lead frame with a wire bond from the top of the chip to a metallized trace on the substrate. Temperature changes cause these various metals to expand and contract at different rates. High temperatures cause reduced lifetime rather than immediate failures. Internal LED junction temperature depends on ambient temperature, power applied to the LED, and thermal resistance from LED chip to ambient.

LED DISPLAYS

Electronic displays include incandescent, liquid crystals, electrochronic, and the light-emitting diode, among others. This section is dedicated to the LED display. Probably the most useful form of the LED is in digital displays. The LED digital display is reliable and inexpensive. It is usually driven from a low-level dc voltage but has comparatively high current draw. Some disadvantages are that under high-intensity ambient light they tend to look faded. As you would understand, large displays are expensive, since they would require large quantities of GaAsP material. Manufacturers such as Litronix have developed light-pipe construction to enhance the reflective light. This allows the display to be seen for distances up to 10 feet. The process involves the placement of a reflecting light pipe over the LED chip. The LED emits light that is reflected off the light-pipe cavity walls and emerges as a larger lighting surface. Light pipes are often filled with glass particles to enhance reflection.

LED displays come in a variety of sizes and colors. The most popular heights are 0.3 and 0.6 inch. Character sizes usually run up to 1.0 inch. Displays come in one-, two-, three-, and four-digit. However, they are not limited to number. Colors are variations of red, yellow, and green.

Seven-Segment Displays   Figure 6–15 is an illustration of typical displays. The displays are called seven-segment because the numbers on the display are made up of seven diffused segments. There are, of course, other types of display fonts such as the nine- and sixteen-segment fonts and the dot matrix fonts.

Figure 6–15   Red Reflector Displays (*Courtesy Litronix, Inc.*)

Each font has special uses such as numeric, hexidecimal, alphabetic, alphanumeric, and other characters. It is beyond the scope of this book to cover all these in detail. Therefore we shall limit our discussion to the seven-segment display.

Figure 6–15 shows a number 4 with a decimal point on its left, a number 8 with a decimal point on its right, a single number 9 with no decimal point, and a multidigit number 6.283 with the decimal point between the 6 and the 2. These are representative of the seven-segment displays available. Figure 6–16 shows characteristic curves and pin designations for a 0.3-inch seven-segment numeric display. The top two drawings are pin layouts. The displays are each 14 pin.

The display on the left has a decimal on its right while the right display is vice versa. Note also that the left display is a common cathode while the right display uses a common anode configuration. The clockwise arrangement of segments begins with segment A on the top and moves clockwise around the display with segment G in the center. This is standard for LEDs throughout industry. Below the display pin layouts is a set of curves for the different-color LEDs. These curves represent the relative intensity versus wavelength of the primary colors green, yellow, high-efficiency red, and GaAsP red. These of course, are the basic LED wavelengths. The lower four curve plots are forward current versus forward voltage characteristics for each of the four colors: DL (high red), DLO (GaAsP red), DLY (yellow), and DLG (green).

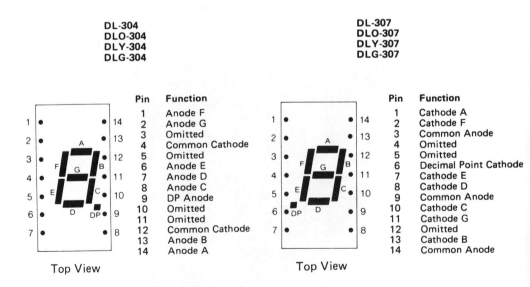

Figure 6–16 Characteristic Curves for 7-Segment Display (*Courtesy Litronix, Inc.*)

Figure 6–16 *continued*

LED Driver Circuitry   There are two methods of driving LED displays. These are the direct dc drive through a decoder/driver of a single LED display and the multiplexing (strobing) of a multiple number of LED digits. Figures 6–17, and 6–18, are representative of the direct dc drive through a decoder/driver.

PIN CONFIGURATIONS

DL-707, DL-707R

PIN	FUNCTION
1	CATHODE A
2	CATHODE F
3	ANODE E,F,G (707 DP)
4	NC
5	NC
6	DECIMAL POINT CATHODE
7	CATHODE E
8	CATHODE D
9	ANODE C,D (707R DP)
10	CATHODE C
11	CATHODE G
12	NC
13	CATHODE B
14	ANODE A,B

Jumper Pins 3, 9, and 14 on Circuit
Board for Common Anode

TYPICAL DRIVE CIRCUITRY

DL-707, DL-707R

Figure 6–17 Direct DC Single-Diode LED-Drive Circuitry (*Courtesy Litronix, Inc.*)

In figure 6–17, the pin configuration of a 0.3-inch seven-segment common-anode red numerical display. This is a simple method of driving the segments. The digit has its own decoder/driver. Switching outputs from the decoder control current flow to each LED through current-limiting resistors. The display and the decoder/driver are TTL matched. Decimal logic is provided along with input and output blanking functions (RBI and RBO).

In figure 6–18, the typical drive circuitry of a 0.6-inch seven-segment series LED display is shown. While the two diodes in series may cover a larger viewing area, the source current from a drop in supply voltage V_{CC} may not illuminate the LEDs sufficiently. As in the standard dc driver circuit with a single LED, a decimal-point driver is supplied.

PIN CONFIGURATIONS

Figure 6–18 Direct DC Double-Diode LED-Drive Circuitry (*Courtesy Litronix, Inc.*)

MULTIPLEXING LED DISPLAYS

The second method of driving seven-segment displays is by multiplexing (strobing). Driving a single-digit display with its own decoder/driver is the simplest method, but does not carry the same efficiency as the multiplexed display. The costs are also a factor, for a large reduction in parts and wiring means cost effectiveness. Furthermore, the reduction in current requirements may even mean a smaller power supply.

Figure 6–19 is the drive circuitry for a multiplexed eight-digit LED display. BCD data are input to a decoder/driver. This particular driver is a TTL logic device as are the rest of the components. The corresponding cathodes of each digit are bussed together. The buses each have their individual current-limiting resistors. The display data are presented to the decoder driver together with an enable signal to the appropriate anode driver. Each digit anode is driven by a switch capable of passing the full current to each display's segments. A simple switch is the PNP high-current or amplifier transistor. In operation the anode switches are operated one at a time in the desired sequence while the appropriate digital data are presented to the decoder/driver. The digit-select multiplexing logic enables each anode driver in a time-sharing process.

Figure 6–19 A Multiplexed LED Display (*Courtesy Litronix, Inc.*)

A number of factors must be taken into account when deciding on the design of a multiplexed display. Besides the optical output, thermal considerations are very important. Most quarter-inch size LED characters are rated at 30 milliamperes dc max per segment. Under pulsed operation, higher currents can be used provided several thermal considerations are taken into account. The average power dissipation must not exceed the maximum rated power. The power pulse width must be short enough to prevent the junction from overheating during the pulse. This implies that the pulse width must get shorter as the amplitude increases.

Present experience indicates that for pulses of 10 microseconds the amplitude should be limited to 100 milliamperes max. Shorter pulses of higher amplitude may be used, but the circuit problems become severe if the pulse width is very short.

THE MONOLITHIC DISPLAY

The monolithic (dual in-line package) display is different from other displays in that all the segments made of light-emitting diodes are diffused on a single chip (see figure 6–20). The chip is usually made from GaAsP as a substrate of N-doped material, while the junction is P+ diffused. The operating wavelength of the monolithic DIP is around 655 nanometers. Since the price of GaAsP is quite high, the monolithic chip size is kept small. An external lens is used to magnify the size of characters. The actual heights of the character are much smaller than the apparent size of the digits.

The lens is placed on the chip by two methods, immersion and nonimmersion. The immersion method molds the lens directly to the chip in cast epoxy. An aluminum contact for each digit segment is wire bonded to an anode contact. The chips are attached to a ceramic or lead frame where the common cathode electrical connection is formed.

The nonimmersion lens differs from the immersion lens in that there is a layer of air between the chip and the lens. Nonimmersion lenses are constructed by attaching the chip to a circuit board. Board size is dependent on the number of display digits. Aluminum contacts for each LED segment are wire bonded to the anode traces on the printed circuit board. Precision-molded lenses are aligned and attached to the printed-circuit board. A second lens is usually attached to add height to the characters.

Figure 6–21 is a pinout-and-dimension illustration of typical red monolithic DIP displays. Figure 6–21A represents a two-digit cast lens display. It is common-cathode multiplexed. As you can see, it has characters that are 0.15 inches in height and 0.11 inch wide. Figure 6–21B shows three- and four-digit

Figure 6–20   Red Multidigit Magnified Monolithic DIP (*Courtesy Litronix, Inc.*)

encapsulated lens displays. These are also common-cathode multiplexed. Both
these displays have characters that are 0.11 inches in height and 0.08 inches wide.

As you can see in the photograph in figure 6–20, the monolithic DIP is
extremely small. They are also low cost and low power. The monolithic DIP has
a wide viewing angle that allows for maximum outdoor visibility. These displays
are used for hand calculators and digital watches.

INTELLIGENT DISPLAYS

Intelligent displays are compact modular components that convert computer data
to readable numbers and letters. They combine CMOS ICs, hybrid construction,
and plastic immersion optic display technology as a single product to simplify the
work of the designer.

DL-44M

PIN	FUNCTION	PIN	FUNCTION
1	NC	7	CATHODE DIGIT NO. 2
2	ANODE E	8	ANODE B
3	ANODE D	9	ANODE G
4	OMITTED	10	ANODE A
5	ANODE C	11	ANODE F
6	D.P. ANODE	12	CATHODE DIGIT NO. 1

(a)

DL-330M

Pin	Function
1	Cathode D1
2	Anode E
3	Anode D
4	Cathode D2
5	Anode C
6	Anode DP
7	Cathode D3
8	Anode B
9	Anode G
10	Anode A
11	Anode F
12	No Pin

(b)

DL-340M

Pin	Function
1	No Connection
2	Anode E
3	Anode D
4	Anode C
5	Anode D.P.
6	Anode G
7	Cathode 4
8	No Pin
9	Anode B
10	Cathode 3
11	Anode F
12	Cathode 2
13	Anode A
14	Cathode 1

(c)

Figure 6–21 Pinout and Dimension Outline of Red Multidigit Monolithic DIPs (*Courtesy Litronix, Inc.*)

105

Construction   The basic intelligent display is constructed with a four-layer hybrid base. The CMOS IC is installed underneath the dual in-line package (DIP). The LED-segmented chip is diffused on a single chip on top of the DIP. This unit is a four-digit display. The LED chip is usually made from GaAsP. The wide-angle lens and cap are installed on the DIP by the immersion method, which molds the lens directly to the chip in cast epoxy. An aluminum contact for each digit segment is wirebonded to common contacts (see figure 6–22).

Figure 6–22   Construction of the Intelligent Display (*Courtesy Litronix, Inc.*)

Electrical Description   Under normal usage, a display would have traditional problems of hookup. These are segment decoding, driving, and multiplexing. The intelligent display eliminates these design problems. In May of 1977 Litronix Inc. introduced the first of the intelligent displays. As a system, the intelligent display is a single device that takes alphanumeric signals from a microprocessor data bus, stores them, then translates and displays the signals as numbers and letters.

Figure 6–23 is an internal block diagram of the Litronix DL 1416 alphanumeric intelligence display. The unit consists of a monolithic LED display and a single integrated-circuit chip. The display is a 4-character, 16-segment alphanumeric monolithic LED, mounted on a 1.2 × 1.0 inch PC card. The LED is magnified to a height of 160 mils. Behind the 4-digit display is a complementary-metal-oxide semiconductor (CMOS). The CMOS chip is large-scale integration (LSI). It contains the necessary logic to translate 7-bit data from a microprocessor into any of 64 alphanumerical characters. Functional components within the CMOS

Figure 6–23 Internal Schematic of an Alphanumeric Intelligence Display (*Courtesy Litronix, Inc.*)

LSI chip are 16 segment drivers, 4 digit drivers, 64-character ROM, 4-word, 7 bit RAM, an internal oscillator for multiplexing, a multiplex counter/decoder, cursor RAM, write address decoder, and level shifters for the inputs.

Figure 6–24 is the pin layout of the alphanumeric intelligent display. The inputs to the chip are as follows:

$\overline{\text{CE}}$ Chip Enable (active low)

This determines which device in an array will actually execute the loading of data. When the chip enable is in the high state, all inputs are inhibited.

A_0, A_1 Digit Address

The address to the display determines the digit in which the data will be written. Address order is right-to-left for positive-true address.

D_0–D_6 Data Lines

The 7 data input lines are designed to accept the 64-character code set.

$\overline{W}$ Write (active low)

Data to be written into the display must be present before the leading edge of write. The data and address must be stable until after the trailing edge.

$\overline{CU}$ Cursor (active low)

When the CU is held low, the display enables the user to write or remove a cursor in any digit position. The cursor function lights all 16 segments in the selected digits without erasing the data. After the cursor is removed, the digit will again display the previously written character.

V+ Positive Supply

TTL compatible +5 volts

V− Negative Supply

Ground

Pin		Function	Pin	Function	
1	D5	Data Input	11	A1	Digit Select
2	D4	Data Input	12	Unused	
3	DØ	Data Input	13	Unused	
4	D1	Data Input	14	Unused	
5	D2	Data Input	15	Unused	
6	D3	Data Input	16	Unused	
7	CE	Chip Enable	17	Unused	
8	W	Write	18	V+	
9	CU	Cursor Input	19	V−	
10	AØ	Digit Select	20	D6	Data Input

Figure 6–24 Alphanumeric Intelligent Display PIN Layout *(Courtesy Litronix, Inc.)*

Figure 6–25 illustrates the 64-character set along with loading data and loading cursor. Data entry is asynchronous and random access. Each digit will continue to display the character last written until replaced by another. The cursor function causes all 16 segments of a character to turn on. The cursor is not a character and upon removal leaves the character unchanged. The cursor is loaded as follows: $\overline{CE}$ and $\overline{CU}$ are held low. A write ($\overline{W}$) pulse will now store the cursor in all digit positions for which the first four data lines (D0, D1, D2, D3) are held high. If previously stored, the cursors can only be removed if their respective data lines are held low while $\overline{CE}$, $\overline{CU}$ are low and write ($\overline{W}$) occurs.

CHARACTER SET

NOTE: All undefined data codes that are loaded or occur on power-up will cause a blank display state.

LOADING DATA

CE	CU	W	A₁	A₀	D6	D5	D4	D3	D2	D1	D0	DIGIT 3	DIGIT 2	DIGIT 1	DIGIT 0
H	X	X	X	X	X	X	X	X	X	X	X	NO CHANGE	NO CHANGE	NO CHANGE	NO CHANGE
L	H	L	L	L	H	L	L	L	L	L	H	NO CHANGE	NO CHANGE	NO CHANGE	A
L	H	L	L	H	H	L	L	L	L	H	L	NO CHANGE	NO CHANGE	B	A
L	H	L	H	L	H	L	L	L	L	H	H	NO CHANGE	C	B	A
L	H	L	H	H	H	L	L	L	H	L	L	D	C	B	A
L	H	L	L	L	H	L	L	L	H	L	H	D	C	B	E
L	H	L	H	L	H	L	L	H	L	H	H	D	K	B	E
L	H	L	–	–	–	–	–	–	–	–	–	SEE CHARACTER SET			

X = DON'T CARE

LOADING CURSOR

CE	CU	W	A₁	A₀	D6	D5	D4	D3	D2	D1	D0	DIGIT 3	DIGIT 2	DIGIT 1	DIGIT 0
H	X	X	X	X	X	X	X	X	X	X	X	D	K	B	E
L	L	L	X	X	X	X	X	L	L	L	H	D	K	B	▨
L	L	L	X	X	X	X	X	L	L	H	L	D	K	B	E
L	L	L	X	X	X	X	X	L	L	H	L	D	K	▨	E
L	L	L	X	X	X	X	X	L	H	L	L	D	▨	B	E
L	L	L	X	X	X	X	X	H	L	L	L	▨	K	B	E
L	L	L	X	X	X	X	X	H	H	H	H	▨	▨	▨	▨
L	L	L	X	X	X	X	X	L	L	L	L	D	K	B	E

X = DON'T CARE

Figure 6–25 Character Set for Litronix DL-1416 Intelligent Display (*Courtesy Litronix, Inc.*)

A "lamp test" function is realized by simply storing a cursor in all four digits of a display. All undefined data characters stored in memory will override a cursor display.

Data is loaded as follows: The chip enable ($\overline{CE}$) held low will enable data loading. The desired data code (D0–D6) and selected digit address (A_0, A_1) should be held stable while write ($\overline{W}$) is low for storing new data. Data entry may be asynchronous and random order.

A Simple System Hookup   A simple 16-digit display system can be set up in an extremely easy manner. Data lines, address lines, and digit select lines along with ± dc power are connected. The significant part of this simple hookup is that only the decoder and 4 four-digit displays are required along with microprocessor inputs. Simplicity in system connection is the intelligent displays greatest asset (see figure 6–26).

Figure 6–26   A Simple System Hookup (*Courtesy Litronix, Inc.*)

Litronix Intelligent Displays   Figure 6–27 illustrates three types of Litronix intelligent displays. Figure 6–28 provides a dimensional drawing for one of the displays, the model DL–1416. Optical specifications for model DL–1416 include luminous intensity, viewing angle, digit size, spectral peak wavelength, and spectral line half-width.

INJECTION LASER DIODE (ILD)

The semiconductor injection laser diode (ILD) is extremely well suited for use within the fiber-optic industry. The prime reasons for their applicability are their inherent ruggedness, extreme efficiency, and small size. They can be pumped and

Figure 6–27 Litronix Intelligent Displays (*Courtesy Litronix, Inc.*)

NOTE: Dimensions in inches ± 0.005 unless otherwise specified

Figure 6–28 Dimensional Drawing of the Litronix Model DL-1416 Intelligent Display (*Courtesy Litronix, Inc.*)

Figure 6–29 The T-912 AlGaAs Injection Laser Diode (*Courtesy ITT Electro-Optical Products Division*)

modulated by injection current. Figure 6–29 is an ITT Model T912 double-heterojunction AlGaAs injection laser typical of the present-day state-of-the-art ILD's on the market. The term heterojunction requires some explanation. Early in the development of semiconductors, lasing action by stimulated recombination of carriers injected across a PN junction was predicted. In 1962, lasing action was achieved in a crystal of gallium arsenide (GaAs) (see figure 6–30). It was accomplished by three independent research teams simultaneously. The firms were GE, IBM, and Lincoln Laboratory. The unit was called a homostructure PN-junction laser. The word homostructure is used because the injection laser diode was made from one material, gallium arsenide (GaAs). The lasers are called junction laser diodes, junction lasers, and injection laser diodes because electrons are injected into the junction region. The gallium arsenide (GaAs) is lightly doped with suitable impurities to form a PN junction just as silicon is doped when manufacturing solid-state diodes for use as rectifiers. The ends of the GaAs crystal are polished to a mirrorlike finish. Light is generated by injecting current (electrons) into the P-type region where there are deficiencies of free electrons in its lattice structure. The N-type region has an excess of free electrons. When the electrons and holes recombine within the junction region (active region), photons are emitted as radiant energy. The mirror ends tend to reflect the photons back into the active (more recombinations) and serve as a feedback mechanism. The sides of the device are optically diffused. Brightness is controlled by adjusting the current flow. As noted, the ILD is placed in a circuit in reverse of a standard diode because current (electrons) are injected into the P-type material.

The homostructure diode quickly fell into obscurity for it was found that the ILD could be improved considerably with the use of the heterostructure. In the heterojunction, the active region is bounded by wider band-gap regions.

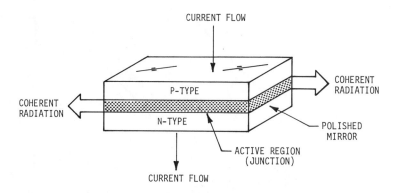

Figure 6–30 A Homostructure PN Junction Laser (*Mike Anaya*)

Heterojunctions fall into three categories, the single heterostructure (SH), the double heterostructure (DH), and the separate confinement heterostructure (SCH). The single heterostructure has only one heterojunction so that injected carriers are confined by the junction at only one boundary of the active area. The double heterostructure (DH) laser carriers and waveguide have boundaries on both sides of the active region. In the separate confinement heterostructure (SCH), the carriers are confined in a region within the waveguide.

A cross section of the T912 double heterojunction injection laser diode is illustrated in figure 6–31. Note that the active area is aluminum gallium arsenide (AlGaAs) while one boundary to the active region is N-type AlGaAs and the opposite boundary to the active region is P-type AlGaAs. On the top of the structure is a layer of P-type gallium arsenide (GaAs), while on the base is an N-type gallium arsenide (GaAs) substrate. The T–912 device is of a stripe geometrical design, that is, a layer of silicon dioxide (SiO_2) is placed between the junctions and a copper. The purpose of the SiO_2 is to provide a narrow stripe rather than a wide structure that may eliminate noise.

The output characteristics of the T–910 are compatible with PIN and APD detectors and photocathodes. They are ideally suited for use as light sources in high-performance, long-distance, high-bandwidth fiber optic communication systems. They may also be used as target designators, intrusion alarms, and other applications.

A dimensional outline of the T–912 is illustrated along with some of its characteristic curves. The edge-emitting chip is mounted P-type side down on a copper sink, which also provides an on-axis alignment with the coaxial stud header (see figure 6–32).

SiO$_2$

p GaAs

p AlGaAs

AlGaAs
(ACTIVE
LAYER)

n AlGaAs

CLEAVED
END MIRROR

n GaAs
(SUBSTRATE)

METAL

METAL

Figure 6–31 A Heterostructure PN Junction Laser (*Courtesy ITT Electro-Optical Products Division*)

PHOTOTRANSISTOR

The importance of the phototransistor has diminished somewhat in the last several years because of the great strides in other optoelectronic devices such as the detector diode. The one important thing that a phototransistor has that other devices lack is the ability to detect light and to amplify within a single device. This is why the phototransistor is still utilized in applications such as position detectors, intrusion alarm sensors, and optical tachometers. In operation, the phototransistor has a large collector-base junction where incident light (photons) can generate hole pairs. The forward-biased base causes light-induced collector current. Operation of the phototransistor is the same as any transistor with photons of light causing the current flow. The phototransistor may not require external bias unless specific transistor operating levels are demanded.

Figure 6–32 Characteristic Curves of an AlGaAs Injection Laser Diode (ILD) (*Courtesy ITT Electro-Optical Products Division*)

Figure 6–33 shows a variety of phototransistors. The shape and size are dependent on the applications. Phototransistors are made from silicon or germanium, much the same as ordinary transistors. They are constructed in the NPN or PNP configuration. Phototransistors are subject to the typical problem of all transistors, temperature variations. These problems may be resolved by biasing techniques and thermal-stability resistors.

Figure 6–33   Phototransistors (*Courtesy Litronix, Inc.*)

Figure 6–34 shows some typical applications. In the figure there are two applications. The first circuit utilizes a phototransistor with a load resistance in its emitter. Output of the transistor is taken from across the load. The second application uses a phototransistor to turn on and turn off data. The output of the transistor is fed through a pair of inverters.

OPTO-ISOLATORS

The opto-isolator is a device with a light source coupled to a light sensor. It transmits while maintaining a high degree of isolation between its input and output. Previously this job was accomplished by relays, isolation transformers, and block-

Figure 6–34 Phototransistor Applications (*Courtesy Litronix, Inc.*)

ing capacitors. The opto-isolator replaces these devices and adds better reliability and function. There are three basic opto-isolator types (See figure 6–35). These are the LED-photodiode, the LED-phototransistor, and the LED-photo-Darlington. In each of these cases the LED is the electroluminescent device, whereas the diode and transistors are the detectors. Input leads are supplied from a current source, while output leads may be tied to a variety of electronic circuits, which we shall discuss later on in this section.

LED – Photodiode.   **LED – Phototransistor**

(a)   (b)

LED – Photo-Darlington

(c)

Figure 6–35 Opto-Isolator Types (*Courtesy Litronix, Inc.*)

With no external optic characteristics, opto-isolators are specified as electrical devices. There are functionally two parameters that define the opto-isolator. These are how well they transfer information from input to output and how efficiently they maintain electrical isolation from input to output. Transfer efficiency is the ratio of output current to input current expressed as a decimal. The current transfer ratio (CTR) varies with temperature and input current.

Cross Sections of an Opto-Isolator   A basic opto-isolator consists of a gallium arsenide (GaAs) infrared diode (IR LED) and a silicon phototransistor mounted together in a dual in-line package. When forward current is passed through the IR LED, it emits infrared radiation at about 900 nanometers wavelength. This radiant energy is transmitted through an optical coupling medium and falls on the base surface of the phototransistor. Phototransistors are designed to have a very large base region and therefore a very large base-collector junction area with a small emitter area. Some incident energy in the form of photons causes the formation of electron-hole pairs in the base region.

Referring to figure 6–36A, the emitter and the base are grounded. A positive voltage is applied to the collector. This enables the device as a photo diode. The high field across the collector-base junction quickly draws the electrons across into the collector region. The holes drift toward the base terminal and attract electrons from the terminal. It is in this manner that electron current flows between the base and collector causing a voltage drop across the load resistor (R_L). A time constant is developed with the junction capacitance (C_{cb}) and the load resistor (R_L), with a corresponding output voltage rise time. The output current in this configuration is quite small and therefore this configuration is seldom used.

The most common lashup is the open base connection as in figure 6–36B.

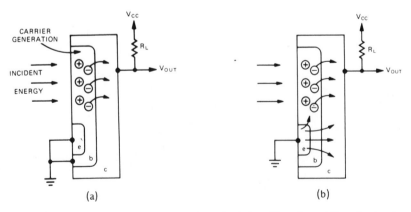

Figure 6–36   Cross Section of an Opto-Isolator (*Courtesy Litronix, Inc.*)

In this configuration, the base is left open and the emitter is grounded. Holes generated in the base region cause the base potential to rise, forward biasing the base-emitter junction. Electrons are injected into the base from the emitter to neutralize the excess holes. Because of the close proximity of the collector, the probability of electrons recombining with these holes is remote and most of the injected electrons are immediately drawn into the collector region. Collector current increases as in any forward-bias transistor, depending on the current gain (beta) of the transistor. Current gain for the B configuration is several hundred times greater than in the A configuration. The current gain, however, is beta times as great as in the A configuration, which causes the rise time to increase by a factor of beta.

Increasing Opto-Isolator Speed The speed at which an opto-isolator responds to a signal depends almost entirely on the detector (sensor). Switching time is a function of the base storage time and the RC time constant of the output. The larger the collector-base region, the more sensitive the sensor and the larger the open circuit collector-base capacitance. This results in long storage time and slow switching time. In schematic A of figure 6–37, a base-to-emitter bleed resistor is added. The bleed resistor provides an external path for removal of the stored charge. This decreases the phototransistor's turn-off time, but reduces sensitivity and current transfer ratio (CTR).

In schematic B of figure 6–37, a cascode amplifier is illustrated. In this case the phototransistor is the current source and the effective load resistance R_L is made up of the cascoded isolation amplifier emitter resistance. This emitter resistance may be only a few ohms. The cascoded transistor is chosen with a low open-circuit capacitance.

In schematic C of figure 6–37, an operational amplifier is placed in the output circuit. The phototransistor is used as a current source driving the summing point of a general-purpose operational amplifier. This has the same effect as the cascode circuit. It lowers the effective R_L with resulting higher speeds.

Circuit Applications Using Opto-Isolators Opto-isolators have a select group of applications. Although most designers are well aware of their distinct capability, there may be a reluctance to utilize them in everyday design. There are three major categories in which the opto-isolator is a standout device because of its unique properties: level translating circuits, remotely actuated switching circuits, and circuits requiring a high degree of electrical isolation. The last category is the most obvious. A prime application of this type would be the electrocardiograph amplifier, which provides the high degree of isolation necessary to ensure patient safety in medical electronics. A second use for an opto-isolator is with an

Figure 6–37   Increasing Speed of Opto-Isolator Response (*Courtesy Litronix, Inc.*)

isolated voltage operated potentiometer. The opto-isolator makes the potentiometer totally independent of the control-voltage reference supply. In remotely actuated switching circuits, the opto-isolator is used as a polar switch. The opto-isolator can also be used as class B amplifier, a convenient means of inputting data to a calculator chip, and as a high-voltage series regulator. Let's look at several basic circuits to see their hookups. One of the simplest uses that can be made of opto-isolators is with transistor-transistor logic (TTL). There are two ways that TTL can be utilized, either driven by or as a driver for the opto-isolator. The matching of the two devices can be handled quite easily within the parameters of the devices.

Figure 6–38 illustrates opto-isolators being used as output driver circuits. In illustration A, two TTL active low circuits are suggested. In the upper circuit, a gate is driven by an opto-isolator with a 12 kΩ pull-up resistor. This circuit is recommended for DTL and TTL digital interfaces. In the lower circuit, a gate is driven by an opto-isolator with a 3 kΩ pull-up resistor. Figure 6–38B illustrates two TTL active high circuits. Both circuits utilize OR gates. The upper circuit is recommended if both positive and negative power supplies are available. The lower circuit is suggested if there is no negative supply.

Figure 6–38 TTL Logic Being Driven by Opto-Isolators (*Courtesy Litronix, Inc.*)

Figure 6–39 represents an opto-isolator used in a logic operation. In the illustration, a pair of opto-isolators are used to perform logical OR functions. This circuit is equally compatible with DTL or TTL logic families. There are two ways in which TTL may be used as a driver for the opto-isolator (see figure 6–40). In the A illustration the TTL is used in an active level high or a current source mode. The output of the TTL is placed on the anode of the LED. In the B illustration, the TTL is used as an active level low or a current sink mode. The output of the TTL is placed on the cathode of the LED. Care must be taken when using logic gates with opto-isolators that the logic mathematical expression be maintained. A high out of a gate used as a current source would turn the LED on. A high out of the same type gate used as a current sink would turn the LED off.

Figure 6–41 illustrates three examples of linear circuit applications. The first utilizes a differential amplifier. In the A circuit the LED must be forward-biased to a suitable current flow. Modulating signals may then be impressed on

Figure 6–39 Opto-Isolator Used to Perform a Logic Function (*Courtesy Litronix, Inc.*)

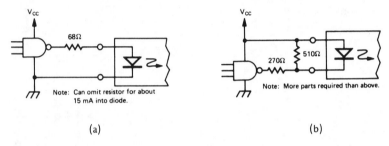

(a)   (b)

Figure 6–40 TTL Logic Driving Opto-Isolators (*Courtesy Litronix, Inc.*)

(a)   (b)

(c)

Figure 6–41 Linear Circuit Applications (*Courtesy Litronix, Inc.*)

that dc bias. In the B circuit a feedback amplifier is utilized to lower the impedance of the input from the opto-isolator. In the C circuit a high-speed operational amplifier is used to provide lower effective load resistance, therefore higher speeds. To repeat a previous note, the opto-isolator has two basic requirements to live up to, efficient transfer of information from input to output and the maintenance of electrical isolation from input to output.

Typical Opto-Isolators   Typical of the opto-isolators are the Litronix IL–1 and the ILQ series (see figure 6–42). The IL–1 is a 6-pin device while the ILQ is a 16-pin device. The IL–1 houses a single opto-isolator. The ILQ houses 4 separate opto-isolators. A connection diagram is provided in figure 6–43.

Figure 6–42   Opto-Isolators (*Courtesy Litronix, Inc.*)

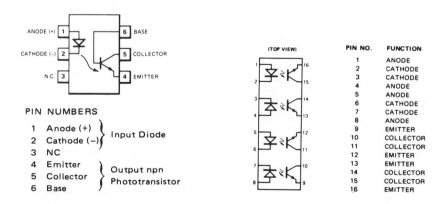

Figure 6–43   PIN Connections for Litronix IL-1 and ILQ Opto-Isolators (*Courtesy Litronix, Inc.*)

LASER OPTICS

Since most laser technology deals in some way with optics, it seems appropriate to take at least a cursory look at some of these devices. Optics are used throughout the laser industry in various applications. Some of these are as mirrors, reflectors, filters, collimators, polarizers, beamsplitters, and magnification. Each time an optic is used, consideration must be made as to its size and application. Optics are coated or uncoated, ground or unground, high or low quality. They may be required to withstand high temperatures, and may be manufactured singularly or in mass production using a variety of materials from glass to germanium.

Optic Geometrics   Figures 6–44 and 6–45 illustrate the geometry of some of the available optics on the markets. Dimensions usually include the length, height, thickness, and/or diameter. Other parameters are the angle at which they are mounted, the angle from the front to the back diameters, and the focal length.

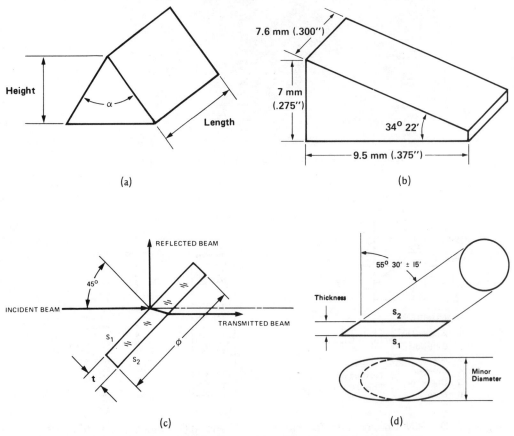

Figure 6–44   Unmounted Optic Geometrics *(Courtesy Coherent, Inc.)*

(a)

(b)

(c)

(d)

Figure 6–45 Mounted and Unmounted Optic Geometrics (*Courtesy Coherent, Inc.*)

Specifications also include whether they come mounted or unmounted and if the mount is or is not part of its calibration or alignment. During design, the engineer may choose an off-the-shelf optic or demand a specially developed lens for a specific job. In any event, the choice of optics that is available, along with the designs that can be made, provide the user with an unlimited catalog for a variety of uses.

The General-Purpose Window   The general-purpose window is chosen for its high light transmission and its surface qualities. For uses in ultraviolet to near infrared wavelength ranges, windows are made from fused silica. Infrared red windows are made from zinc selenide. The windows are flat with slight wedges of about 10 arc sec angle. A wedge is the small angle from the front to the back side of the mirror on its perimeter. Specifications for the window include the diameter of the front and back surfaces, the thickness of the window, and whether or not the surface is antireflection coated. Surfaces are polished to provide low scattering and to prevent etalon effects. Etalon effects indicate a change in optical path differences due to a mechanical displacement or a difference of refractive index (see figure 6–46).

Substrate Specifications

Material: Schlieren Grade
Fused Silica

Surfaces: S_1 Plano
$\lambda/10$ 10–5 Laser quality
S_2 Plano
$\lambda/10$ 10–5 Laser quality

Thickness: 2mm ± .1mm

Diameter
Tolerance: +.00mm −.25mm

Wedge: <10 arc seconds

Figure 6–46 General Purpose Windows (*Courtesy Spectra-Physics, Inc.*)

Brewster-Angle Windows   The Brewster window (see figure 6–47) is a window placed on the ends of a plasma tube at an angle that allows maximum light power to pass through the active medium, which is usually gas. A plane parallel window would reflect much of the light. Brewster windows are made from top-grade fused silica. The windows are flat with a slight wedge from the front surface to back surface of less than 10 seconds wedge angle. Edges are machined to an elliptical circumference at Brewster's angle, which is around 55°32′. However, this is not a standard and can vary around this angle depending on the manufacturer. Specifications also include the front and rear diameters along with the thickness. As with general-purpose windows the Brewster windows are chosen for their high transmission and surface qualities. Grinding and shaping of the Brewster windows usually takes place prior to polishing. Windows are polished to provide low scattering.

Substrate Specifications

Material:	Schlieren Grade Fused Silica
Surfaces:	S_1 Plano; $\lambda/20$; 10–5; Laser quality S_2 Plano; $\lambda/20$; 10–5; Laser quality
Wedge:	<10 seconds

Figure 6–47   Brewster Angle Windows (*Courtesy Spectra-Physics, Inc.*)

Mirrors   In general, laser mirrors are utilized either external to the laser cavity or within the cavity. The standard mirrors are designed for long life and durability. General-purpose mirrors are made from optical-grade fused silica.

Dielectric metal oxides may be deposited on fused silica substrates to produce laser mirrors that can withstand plasma discharges in laser tubes. The dielectric material must also be able to withstand temperatures at levels between say −260 to +260°C. The specific substrate/coating combination is a matter of wavelength demand. The selection is dependent on the wavelength of the application. Mirror testing includes humidity, water immersion, abrasion, and adhesion tests (see figure 6–48).

Mirrors are shaped according to their use. Specifications include the radii of the first and second surfaces, along with the thickness and diameter of the substrate. A wedge between front and rear surfaces is usually less than a 30-minute angle. Mirrors are chosen to provide low scattering, low temperature distortion, and to maintain high transmission properties. Substrates that are used within a laser cavity must also be durable and are thick to ensure further stability.

R_1 = Radius of first surface
R_2 = Radius of second surface
S_1 = Side one
S_2 = Side two
t = Thickness of substrate
d = Diameter of substrate

Substrate Specifications:

Material: Optical Grade Fused Silica
Surfaces: S_1 λ/20 10–5 Laser quality
S_2 λ/10 10–5 Laser quality
Wedge: less than 30 minutes

Diameter:	7.75mm	15.0mm	25.4mm
Diameter Tolerance:	+.00mm −.20mm	+.05mm −.15mm	+.05mm −.15mm
Edge Thickness Tolerance:	±.3mm	±.5mm	±.5mm
Radius Tolerance:	less than 1% from nominal		
Clear Aperture:	6mm	10mm	15mm

Figure 6–48 Laser Mirror Substrates (*Courtesy Spectra-Physics, Inc.*)

Mirrors usually come in sets that include the high-reflectance (100 percent) mirrors and the partial-reflectance transmission mirrors. In the case of the dye laser, the set would include a folding mirror and a pump mirror.

Front Surface Reflectors Front surface reflectors (see figure 6–49) are primarily designed for beam steering outside the laser. However, they may be used for general-purpose laboratory work not connected with lasers, which require light reflection. Front surface reflectors are made from Pyrex with a dielectric coating. The front surface reflector is polished one side only. They have high reflectivity within the range of designed wavelengths. Operation of the reflector is around ±5 percent about the 45° design angle. Reflection is nominally around 95 percent at design wavelengths. Specifications for the reflector include the diameter of the reflector and its thickness. Other parameters include the angle of incidence in the application (usually 45°) and the device wavelength.

Incident Beam Polarization When optics are used at some angle of incidence other than normal to the surface, the polarization of the incident beam must be taken into consideration (see figure 6–50). A polarized coherent beam is characterized by an electric vector E in a plane perpendicular to the direction of the beam. When the beam strikes a surface a plane of incidence is formed by the beam and the normal to the surface. For the special case where the E vector is perpendicular to this plane the polarization state is called S. If the E vector lies in this plane the polarization is called P. Note that the polarization state depends on the orientation of the component relative to the E vector of the beam. The direction of the E vector of a laser beam is usually determined by Brewster windows or other polarizing elements.

The Beamsplitter The beamsplitter, when placed at a 45° angle of incidence, will reflect a beam from the optic at 90° and transmit a beam through the optic. The reflection/transmission is near 50/50, ±5 percent. The transmitted beam is parallel to the incident beam. Beamsplitters for ultraviolet to near-infrared wavelengths are made from optic-grade fused silica. Infrared beamsplitters are made of polycrystalline germanium. Beamsplitters rear surface has an antireflection coating to minimize surface reflections. All coatings are low loss. The beamsplitters are designed for use at a 45° angle of incidence and for polarization normal (S) or parallel (P) to the plane of incidence. Specifications for the beamsplitter include its diameter and thickness. Other specifications are its wavelength and whether it is polarized normal (S) or parallel (P) to the plane of incidence.

In figure 6–51, transmission curves which relate transmittance in percent to wavelength in nanometers are shown. The two curves represent polarized normal (S) or parallel (P). Transmission for unpolarized light is the average of the (S) and (P) values. The beamsplitter can be applied to beam manipulation, beam deflection, interferometers, imaging, and recording.

Coated Substrate Specifications

Material: Pyrex 7740; fine annealed
Surfaces: S_1 plano; $\lambda/10$; 60–40 per Mil-0-13830A; S_2 fine ground
Edges: As molded
Sizes: 25.4mm diameter $\times$ 12.7mm thick (G0050–000); 50.8mm diameter $\times$ 12.7mm thick (G0051–000)

Coating Specifications

at nominal wavelength:

Angle of incidence: 45°
Normal bandwidth reflectors: $R_{avg} > 99.0\%$ (Typical performance $R_{avg} > 99.5\%$)
Wide bandwidth reflectors: $R_{avg} > 98.0\%$
R_{avg} is the arithmetic average of reflectance in parallel and perpendicular planes of polarization.

Figure 6–49 Front Surface Reflector (*Courtesy Spectra-Physics, Inc.*)

Figure 6–50 Incident Beam Polarization (*Courtesy Coherent, Inc.*)

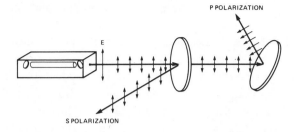

Coated Substrate Specifications

Material: Schlieren grade
fused silica

Surfaces: S₁ plano; λ/20;
40-20 per Mil-0-13830A
S₂ plano; λ/10;
40-20 per Mil-0-13830A

Wedge: 30 min ± 5 min.

Sizes: 25.4mm diameter x 4.2mm thick
(G0020-000)
50.8mm diameter x 8.3mm thick
(G0022-000)

Specific Polarization Control

Coating specifications at nominal wavelength

Angle of incidence: 45°

S₁: Transmission tolerance for "S"
polarization: ±3%

Transmission tolerance for "P"
polarization: ±6%

S₂: Anti-reflection coating
$R_{avg} < 0.6\%$

Useable clear aperture at 45°
incidence equals 53% of diameter

Scatter and absorption loss
< .2% 440nm and higher
< .5% below 440nm

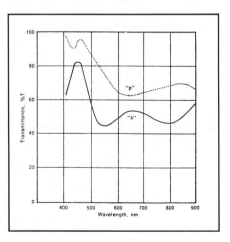

Figure 6–51 The Beamsplitter (*Courtesy Spectra-Physics, Inc.*)

131

Linear Polarizers   Linear polarizers consist of multilayer dielectric coatings deposited on the interface between two cemented prisms. The polarizer coating interacts with randomly polarized light to produce two linearly polarized beams. In figure 6–52A, a linear polarizer is illustrated. In illustration B, S polarization is shown perpendicular to the plane of incidence. P polarization is shown parallel to the plane of incidence. In C, the plane of incidence contains the light ray and the normal to the surface the light ray is falling on. Linear polarizers selectively transmit a P polarized beam, while reflecting an S polarized beam at approximately 40° to the exit beam. The transmitted beam has low wavefront distortion relative to the input beam. The rejected component does not have a controlled wavefront specification. The polarizer coatings are designed for broad spectral ranges, and can be used in both laboratory and production applications.

Specifications for the linear polarizer include the transmittance of the beam, the wavefront distortion, and angular beam deviation.

Specifications

Transmitted beam only

Transmittance: $T > 95\%$ over the wavelength range

Wavefront Distortion: $< \lambda/2$

Angular Beam Deviation: < one minute of arc

Clear Aperture: 5mm diameter, centered

Acceptance Angle: ± 1° over full wavelength range

Anti-reflection coatings: Entrance and exit face $R_{av} < 0.5\%$

Damage Resistance: to 100W/cm² CW; 1 megawatt/cm² peak

Extinction Ration (T_p/T_s): 1000:1

(a)

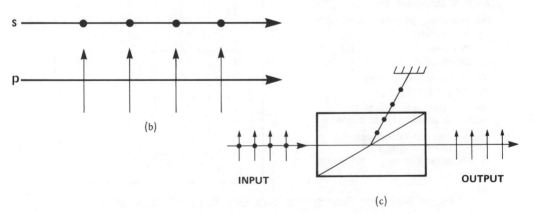

(b)

INPUT   OUTPUT

(c)

Figure 6–52 Linear Polarizer (*Courtesy Spectra-Physics, Inc.*)

Polarization Beamsplitter Cubes Polarization cubes (see figure 6–53) divide an input beam into two linearly polarized beams at right angles to each other. The transmitted beam has the P polarized component and the reflected beam has the S polarized component. Polarized cubes are manufactured from high-quality optical glass to exacting tolerances that maintain low wavefront distortion of the entrance beam. Other specifications include transmittance of the beam, wavefront distortion of the exit beam, and transmitted beam deviation.

Specifications

Transmittance: $T \geqslant 98\%$ of total energy

Wavefront Distortion of Exit Beams: $< \lambda/2$

Reflected Beam: $90° \pm 30$ min from incident beam

Transmitted Beam: Deviation $<$ 5 milliradians

Clear Aperture: 8 mm centered for 12.7mm cube; 16mm centered for 25.4mm cube

Anti-reflection Coating: All entrance and exit faces $R \leqslant 0.25\%$

Damage Resistance: To 100W/cm² CW; 1 megawatt/cm² peak

Extinction Ratio: Specified as minimum in both exit beams; T_p/T_s transmitted beam; T_s/T_p reflected beam

Angle at which maximum extinction occurs: $0 \pm 1°$ incidence

Figure 6–53 Polarization Beamsplitter Cubes (*Courtesy Spectra-Physics, Inc.*)

Applications for the beamsplitter cube are illustrated in figure 6–54. The applications include: (A) isolation of the laser cavity energy from light reflected from an optical system; (B) combination of two beams of the same wavelength but opposite polarizations; (C) read/write equipment; and (D) coupled-cavity experiments in nonlinear optic applications.

Prisms Prisms generally are manufactured in three basic forms: the right angle, the Littrow, and the Brewster angle. These three forms are illustrated in the figure 6–55A, B, and C respectively. The right-angle lens permits general purpose beam bending with good optical figures. The Littrow prism is designed for use as a dispersion element in laser cavities. They are used in applications that

Figure 6–54 Polarization Beamsplitter Cube Applications (*Courtesy Spectra-Physics, Inc.*)

require low scatter. The Brewster-angle prism is also used as a dispersion element in laser cavities. Light enters the prism and is dispersed as it passes through the element and exits at Brewster's angle.

Specifications for prisms are surface quality and angular tolerances. Prisms are made from high-quality optical fused silica. Coating may or may not be desirable as the application demands.

CLEANING LASER MIRRORS

The cleaning of laser mirrors is, as you may have determined, a simple but extremely important task. Mirrors must be free of contamination and dust. The difference between a clean and a dirty optic may be a loss in power output or a change in the beam profile. The procedures listed here are for general maintenance.

Substrate Specifications

Material: BK-7 or equivalent

Surfaces: S_1 Plano $\lambda/10$
40–20 per Mil-O-13830A
S_2 & S_3 Plano $\lambda/8$
40–20 per Mil-O-13830A

Angular Tolerance: $\pm$ 10 minutes

Sides Ground and all edges beveled

Tolerances on dimensions $\pm.25$mm

Substrate Specifications

Material: Schlieren Grade
Fused Silica

Surfaces: S_1 & S_2 $\lambda/20$
10–5 Laser quality
All others ground

Physical Tolerances: $\pm.4$mm

Pyramid Tolerance: ± 5 min.

Size: 11.81mm High
12.70mm Wide
8.07mm Deep

Apex Angle: 34° 28 min $\pm$ 15 min.

Substrate Specifications

Material: Schlieren Grade
Fused Silica

Surfaces: S_1 & S_2 $\lambda/20$
10–5 Laser quality
All others ground

Base Angles: 55° 30 min $\pm$ 15 min

Physical Tolerances: $\pm$ .12mm

Size: 11.46mm High
14.38mm Wide
7.92mm Deep

Figure 6–55 Prisms (*Courtesy Spectra-Physics, Inc.*)

There may be a case where optics have collected contamination beyond normal cleaning maintenance. These may entail the use of more stringent methods such as ultrasonic cleaners and strong detergents. Procedures vary somewhat for different types and manufacturers, but generally follow a procedure such as follows:

1. Select a clean, dust-free atmosphere and a flat working surface.
2. Place a light pencil mark on the top center of the laser mirror before removing from laser head.
3. Remove the laser mirror from the laser head.
4. Place the laser mirror on the flat working surface with the coated side of the optical working surface in the up position.
5. Using a hemostat or other plier-type tool, pick up a lens tissue (folded several times). Do not touch the lens tissue with your fingers.
6. Soak the lens tissue in methanol or acetone. Shake off excess fluid.
7. Place the moist tissue on the coated surface of the lens and make one sweep. Throw away the tissue.
8. Repeat as necessary using a new tissue for each sweep.
9. Place the laser mirror in the laser head aligning with pencil mark made in step 2.

When cleaning filters, etalons, and special lenses, use precautions that are peculiar to the laser. It is advised that manufacturer's precautions be read carefully, for the successful operation power output and beam profile of the laser are dependent on clean optics.

INSPECTION OF OPTICAL SUBSTRATES

In any manufacturing field, inspection is an integral part of the ongoing process. Inspection of optics falls into several major areas. Surface quality is the first of these inspections. The numerical designation of the maximum number and size of surface defects within the clear aperture is defined by MIL–0–13830A. These are referred to as scratch-dig defects and are represented by numbers such as 80–50 and 20–10, etc. Standard quality refers to defects that are visually apparent when illuminated with a standard 40-watt incandescent bulb. Laser quality means those defects that are visually apparent under high-intensity illumination of 10,000 to 15,000 footcandles as obtained from a focused American Optical microscope illuminator or its equivalent. Figure 6–56 illustrates an optic under inspection. A high-intensity illumination ensures conformance to laser quality standards.

Component clear aperture is the next inspection. The area of the polished surface or surfaces of an optical component over which the technical specifications apply is usually defined in linear terms. These terms may be diameter, length, width, thickness, etc. Surface figure tolerance is surface flatness or surface accuracy. The maximum departure of a polished surface from the ideal specified surface is monitored. This tolerance is expressed in wavelengths of light. A typical specification would be 632.8 nanometers wavelength.

Figure 6–56 Inspection of an Optical Substrate under Intense Magnification (*Courtesy Spectra-Physics, Inc.*)

The allowable deviation of the radius of curvatures from the specified value is expressed in units of length. A radius tolerance would be $\pm 1\%$. Edge chips are allowable as long as they do not encroach the clear aperture.

Specifications of polarization are assumed to refer to linearly polarized light in which the electrical (E) vector is parallel, (P) polarization to the plane of incidence, perpendicular, (S) polarization to the plane of incidence, or at an angle of 45° (average polarization) to the plane of incidence. The plane of incidence is defined by the incident beam and a normal to the surface.

THE FLASHLAMP

Introduction (See figure 6–57.) A flashlamp is a gas discharge light source designed to produce pulsed radiation. It is capable of providing much higher peak intensities than are available with continuous light sources. A wide variety of designs exist, each stressing some particular aspect of lamp performance. The range of performance is extremely broad—electrical input energy may be varied from 10^{-2} to 10^5 joules per pulse, pulse length from 10^{-3} microseconds to 10^5 microseconds, and pulse repetition rate from single shot to 5×10^4 Hz.

Radiation Mechanisms It is not simple to characterize the optical output of a flash lamp. The plasma emits radiant energy over an extremely wide range of wavelengths, from far infrared to ultraviolet. The radiation is composed of several different components, each corresponding to a different physical light

Figure 6–57 The Flashlamp In Crossection (*Courtesy ILC Technology*)

emission mechanism. The relative importance of each of these mechanisms depends strongly on the power density in the lamp.

Characterization is made even more difficult by the fact that the plasma is a particularly transparent emitting volume. The power radiated depends on the depth of plasma viewed, and therefore not only on the lamp's dimensions, but also on its orientation.

The discharge in a flashlamp is initiated by causing a spark streamer to form between the electrodes. As the discharge channel grows, the electrical resistance drops sharply. The electrons in the plasma equilibrate in a high temperature distribution very quickly, and ionize and heat the plasma through collision.

FLASHLAMP TRIGGERING

The Triggering Process Ignition of a flashlamp begins with a spark streamer that bridges the gap between the electrodes with a thin, high current density, conductive path through the gas. If the voltage drop across the spark streamer is less than the voltage applied by the energy storage capacitor, current will begin to flow in the main discharge circuit. The spark streamer will then expand in accordance with the discharge circuit parameters. Formation of the spark streamer occurs in two distinct stages. Initially, a spark streamer is formed from one electrode to the inside wall of the lamp. Then the streamer will propagate along the surface toward the other electrode as a sliding spark discharge. The path of the spark is determined by the location of the external trigger wire or ground plane outside the lamp.

Triggering Methods (See figure 6–58.) Three different basic triggering circuits are in common use. Each has its own particular advantages and short-comings, as discussed below.

In external triggering, the high voltage trigger signal is applied directly to the trigger wire. The trigger circuit components can be small, lightweight and inexpensive. The energy discharge circuit is independent of the trigger circuit so that design of the pulse forming inductor is not influenced by triggering considerations. Other trigger methods are more reliable, particularly at low energy storage voltages. Because trigger voltages are exposed, this method is not recommended for humid or high altitude environments.

The series triggering method employs a trigger transformer with its secondary winding in series with the energy storage capacitor and the flashlamp. After flashlamp ignition, the transformer core is saturated by the current flowing in the circuit, and the saturated inductance of the transformer serves as the pulse forming inductor. Absolute triggering reliability can be achieved at low capacitor charging voltages. All high voltages can be encapsulated, providing reliable operation in severe environments. The trigger transformer is generally large, heavy and more costly than other trigger methods. For some circuits, the saturated secondary inductance is too large to comply with other circuit constraints.

In parallel triggering the trigger transformer secondary is in parallel with the flashlamp. Isolation from the energy storage capacitor can be accomplished by a capacitor (in series with the secondary) or a diode train. With a capacitor, the trigger transformer can be small and lightweight, but the circuit then requires a large pulse forming inductor. A diode or diode train avoids this problem, but adds parasitic losses to the circuit.

Figure 6–58 Flashlamp Triggering (*Courtesy ILC Technology*)

Flashlamp Simmer Pulse Operation   (See figure 6–59.) For high repetition rate, high average power applications, simmer-pulse operation of flashlamps may offer substantial advantages.

In simmer-pulse operation a low power DC discharge is maintained in the lamp during the interpulse interval. Pulse energy is gated into the lamp by means of an SCR or switch tube in the energy storage circuit. The flashlamp is initially ionized with an external trigger transformer or the open circuit voltage of the simmer supply.

Simmer-pulse operation is recommended for applications in which the pulse rate exceeds 100 Hz. At high repetition rates there is not sufficient time for the flashlamp to recover its holdoff characteristic, and a series switch in the energy storage circuit will be required.

Figure 6–59 Flashlamp Simmer Pulse Operation (*Courtesy ILC Technology*)

The advantages of simmer-pulse operation are:

a. Higher Nd: YAG pumping efficiency
b. Increased lamp life
c. Improved pulse to pulse reproducibility of flashlamp output
d. Reduced time jitter
e. Higher repetition rate capability

Many of the advantages of simmer-pulse operation can be achieved in small, light weight systems by using the pseudo-simmer technique. The lamp is ignited by a conventional external trigger signal, but initial current flow in the discharge circuit is limited to approximately 50 mA by a series resistance. After a delay of approximately 100 to 200 microseconds, the limiting resistor is shorted out by an SCR switch and the main discharge occurs.

TYPICAL FLASHLAMPS

Typical flashlamps are those illustrated in figures 6–60, 6–61, and 6–62. The first, in figure 6–60 are a group of lamps used in industrial applications. The thin lamp on the top is a krypton flashlamp used in about 75% of all laser resistor trimmers and semiconductor wafer scribers. The center lamp in figure 6–60 is a krypton flashlamp used in high average power industrial lasers that perform a wide variety of welding and drilling operations. The lamp with the cable ends is a xenon flashlamp used in flashlamp pumped dye lasers and in lasers utilizing unstable resonators.

Figure 6–60 Industrial Laser Lamps (*Courtesy ILC Technology*)

In figure 6–61 are lamps for military lasers. All the lamps here are used in military rangefinders and target designators. In the figure 6–62 are lamps used for fusion lasers. These lamps have high voltage (60 kV) insulation systems for use in air and nitrogen filled cavities. Lamp reliability is assured by microscopic inspection during and after manufacture and by severe functional tests.

Figure 6–61   Military Laser Lamps (*Courtesy ILC Technology*)

Figure 6–62   Lamps Used for Fusion Lasers at High Voltages (*Courtesy ILC Technology*)

7

OPTICAL FIBERS, CABLES, AND COUPLINGS

GENERAL

Optical fibers are transparent, dielectric cylinders surrounded by a second transparent dielectric cylinder. The fibers are light waveguides used to transmit energy at optical wavelengths. The light is transported by a series of reflections from wall to wall of an interface between a core (inner cylinder) and its cladding (outer cylinder). The reflections are made possible by a high refractive index of the core material and a lower refractive index of the cladding material. Refractive index is a measure of the fiber's optical density. The abrupt differences in the refractive indexes causes the lightwave to bounce from the core/cladding interface back through the core to its opposite wall (interface). Thus the light is transported from a light source to a light detector on the opposite end of the fiber.

This chapter looks at fibers in detail. We shall discuss types, classes, and manufacturing procedures. We shall make an in-depth study of problems of misalignment and the concepts of alignment. We shall study the fiber specifications and consider the advantages and disadvantages of properties such as dispersion, attenuation, bandwidth, and tensile strength. An informative look at cables will then be made. The final area of coverage in this chapter will be the interface of fibers and cables with couplers and splicing techniques.

Fiber Generalities   The discussion of how light is launched into a fiber and arrives on the other end of the fiber is fairly easy to understand. The details involved in transmission and the mechanics of coupling, however, are manifold. Let us first discuss in lay terms how transmission takes place.

143

Light is propagated through optical fibers in a series of reflections from one side wall to opposite wall. The acceptance angle i must be small enough so that all of the signal is reflected. In figure 7–1, a fiber core is covered by a layer called the *cladding*. The core has a refractive index higher than that of the cladding. Index of refraction is determined by the manufacturer when the fiber is produced. When the light enters the optic fiber, it reflects from wall to wall of the cladding back into the core. The fiber illustrated is a step index fiber. That is, there is an abrupt refractive index change between the core and the cladding. In order that the light beam may bounce freely from wall to wall without loss, the cladding was added. The cladding acts as a mirror to reflect without power loss. A second use of the cladding is to prevent light from escaping to nearby fibers in the event of bundling several together. The cladding also contributes to the strength of the fiber.

$\angle i$ = INPUT HALF ANGLE
N1 = CORE INDEX OF REFRACTION
N2 = CLADDING INDEX OF REFRACTION

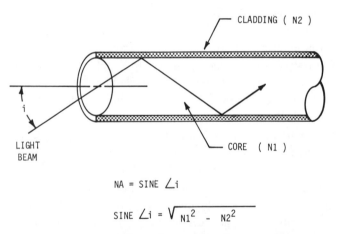

$$NA = SINE \angle i$$

$$SINE \angle i = \sqrt{N1^2 - N2^2}$$

Figure 7–1 Basic Optical Fiber (*Mike Anaya*)

The acceptance angle i is one-half the total cone of acceptance. The cone of acceptance is the light-gathering area at the fiber input. The mechanics of transmission angles are handled in detail later on in this chapter.

Fiber Properties   There are six major properties involved when choosing the correct fiber for an optical system. These are:

1. *Numerical Aperture (NA)*—Numerical aperture is the mathematical since of the angle of acceptance. Half the cone of acceptance is the angle i in figure 7–1.

$$NA = Sine \angle i$$
$$NA = \sqrt{n1^2 - n2^2} = Sine \angle i$$

where   n1 = refractive index of the core

           n2 = refractive index of the cladding

           $\angle i$ = the angle of acceptance

NA is a trigonometric function which represents the sine of the angle i. If the NA increases, the angle i must have also increased and the fiber sees more light. This may seem like a good thing. A problem enters when there is a greater spread in propagated light rays. The increase in ray velocity limits the bandwidth. The effect is called inter-modal dispersion. The NA of a fiber can never be greater than 1.0. This is mathematical maximum sine value. Most NA values are low, between 0.20 and 0.60, and seldom near the maximum sine value. This is a range and must not be judged. For specific values, the manu-facturers' data sheets must be consulted.

2. *Dispersion*—The second property to contend with is *dispersion*. Dis-persion is the spreading or widening of light rays. In fiber optic sys-tems, dispersion is either intermodal or intramodal. The intermodal dispersion (also called *multimode*) is the propagation of rays of the same wavelength along different paths through a fiber. This results in the wavelengths arriving at the opposite end of the fiber at different time periods. The second dispersion, intramodal is determined by three separate factors: material dispersion, waveguide dispersion, and cross-product dispersion. Material dispersion is due to variations of the index of refraction of the core and the cladding materials. Wave-guide dispersion is a phenomena caused by the bandwidth of the signal and the waveguide configuration. Cross-product dispersion is leakage of optical energy from one material, fiber, cable, or connector to other materials. Wavelength and cross-product dispersions are usually small and can be ignored. Dispersion is a function of the refractive index of the material the fiber is made of and the wavelength or mode of the light traveling through the fiber. It is measured in nanoseconds per kilometer (ns/km).

3. *Attenuation*—Attenuation is the loss or reduction in amplitude of transmitted energy. Intrinsic losses are due to light scattering of trans-

mitted energy. Losses of light due to the differences of refractive indexes of two mediums at their interface are called *Fresnel losses*. Intrinsic losses are due to light scattering within the fiber. The most prevalent type of light scattering is Rayleigh. Rayleigh scattering is power loss due to the molecular structure of the fiber material. Other power loss may be due to imperfections or bubbles in the fiber material. Manmade power losses from scratches or dirt are common.

4. *Bandwidth Parameters*—Attenuation curves are always provided by manufacturers of fibers to allow the designer to choose the correct fiber for a particular use. Curves or equivalent tabulated data are expressed in decibels per kilometer (dB/km). The curve in figure 7–2 plots wavelength against decibels. This provides the designer with the fiber's transmission qualities. These parameters are developed during manufacture.

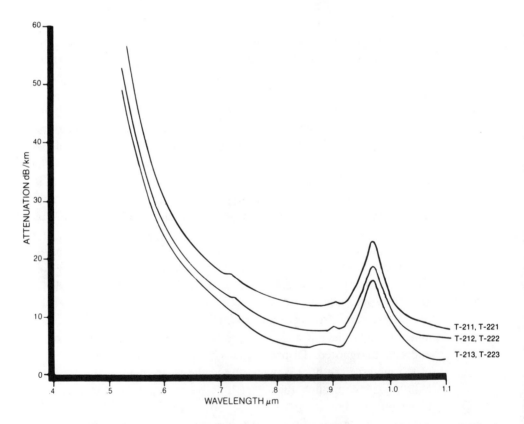

Figure 7–2 Wavelength versus Attenuation Curve (*Courtesy ITT Electro-Optical Products Division*)

TYPICAL SPECTRAL ATTENUATION		
Fiber Type	**Attenuation**	
	@ (85 μm)	@(1.06 μm)
T-211, T-221	12 dB/km	8 dB/km
T-212, T-222	8 dB/km	5 dB/km
T-213, T-223	5 dB/km	3 dB/km
Numerical Aperture	.25	
Fiber Type	**Dispersion**	
	(–3 dB Intermodal)	
T-211, T-212, T-213	2 ns/km	
T-221, T-222, T-223	1 ns/km	

Figure 7–2 *continued* Wavelength versus Attenuation Curve

5. *Rise Time*—The rise time is a parameter that tells the designer that selected parts will operate at speed required. Rise time identifies the fiber's dispersion properties. Rise time is determined by intermodal (multimode) dispersion and/or intramodal (material) dispersion. To find the total rise time of a system you simply add the rise time required for each time-critical component, then add some tolerance. In their technical Note R–5, ITT Electro-Optical Products Division uses this simple calculation. The overall system rise time is 1.1 times the root sum square of all system constituents.

$$T_{SYSTEM} = 1.1 \ \sqrt{T_I^2 + T_M^2 + T_{etc}^2}$$

where T_I = Intermodal rise time

T_M = Material intramodal rise time

6. Fiber Strength—Another fiber property is tensile strength, which is governed by manufacturing intricacies. These techniques eliminate flaws and microcracks in the fiber. Flaw-free fiber cores, claddings, and surfaces are the ultimate attempt in the fabrication of excellent fibers.

It must be noted that these six parameters (numerical aperture, dispersion, optical attenuation, bandwidth and frequency curves, rise time, and fiber strength) are necessary requirements. Additional considerations during design may be size, shape, and bend radius, among others.

FIBER MANUFACTURE

Material Involved in Fiber Manufacture Most fibers are made from fused mixtures of metal oxides, sulfides, or selinides. The optically transparent mixtures used in fibers are called *oxide glasses*. The oxide glasses are common silica (SiO_2); lead silicate, such as crystal; and sodium calcium silicate, such as plate glass. Materials involved in the process would be sodium carbonate, calcium carbonate, boric oxide silica (sand), and lead oxide. The chemical mixtures are extremely complex and must satisfy all the requirements of attenuation, strength, etc. that were just discussed.

General Techniques The manufacturing techniques involved in the production of optic fibers are highly sophisticated processes. Most of these processes are secret or at the least proprietary. Considering the competitive nature of the product this is understandable. There are three general techniques used in the fabrication process. The first to consider is to melt glass in one or two containers and pour the glass together to form the fiber. The second is create a glass core or cladding, then melt a second glass to liquid or glass and deposit the second as a core or cladding on the original. The third is to begin with an exceptional rod of glass and coat it with a plastic cladding. The two processes that we shall discuss are the double-crucible and the chemical-vapor-deposition (CVD) processes.

Double-crucible method of fiber fabrication In early processes, the rod-in-tube method was used (see figure 7–3). This method exudes cladding glass over core glass. The double-crucible furnace is built so that two crucibles holding cladding glass and core glass are vertically aligned. The crucibles are made from platinum or other heat-resistant material. Simple silica crucibles may be used, but they have a tendency to shatter when cooling to room temperature. Each crucible has a narrow aperture at its base. As the cladding glass and core glass are heated, they flow symmetrically through the crucible apertures so that a composite glass flow is produced. The composite glass may be pulled to form a perfect or near-perfect fiber. The dimensions of the two crucibles are very exact, to ensure that core-to-cladding ratios are met. The viscosities and densities of the materials must be considered in flow restrictions, but usually are very similar in both cladding and core. The atmosphere around the crucibles is carefully controlled to prevent contamination and introduction of oxygen, which will cause bubbles in the glass and consequent high scatter losses.

The chemical-vapor-deposition (CVD) process The CVD method was first used by Corning Glass Works of Corning, New York, when they developed the first 20 decibel per kilometer fiber. The technique was called the *soot process*. Figure 7–4 is a line drawing of two doped deposited silica (DDS) fabrication processes. On the left is an inside vapor phase oxidation (IVPO) process. This

CORE GLASS

CLADDING GLASS

FURNACE

GLASS FIBER

ROD-IN-TUBE

DOUBLE-CRUCIBLE

Figure 7–3 Double Crucible Method of Fiber Fabrication (*Courtesy Corning Glass Works*)

SOOT DEPOSITION

DOPED SOOT LAYER
(CORE GLASS)

FUSED SILICA TUBE
(CLADDING GLASS)

INDUCTION REDRAW

SINGLE-MODE OPTICAL
WAVEGUIDE

SOOT DEPOSITION

SiO_2 SOOT LAYER
(CLADDING GLASS)

DOPED SiO_2 ROD
(CORE GLASS)

INDUCTION REDRAW

MULTIMODE OPTICAL
WAVEGUIDE

Figure 7–4 Inside (IVPO) and Outside (OVPO) Vapor Phase Oxidation Processes (*Courtesy Corning Glass Works*)

process is used to create single-mode optical waveguides. The outer layer of glass is a fused silica tube used as the cladding. Gaseous components react in a flame to produce a fine deposit of doped or undoped silica. The soot deposition is internal in this case. It is deposited as a doped layer to form the core. The redraw process involves heating the glass by induction and drawing it to form the single mode optical waveguide. On the right in figure 7–4, the outside vapor phase oxidation (OVPO) process is illustrated. The inner layer provides the core glass. The outer soot layer (that was deposited), establishes the cladding. The redraw process involves heating the glass by induction and drawing it to form the multi-mode optical waveguide.

The doped-deposited-silica (DDS) process shown in the figure 7–5 allows the introduction of carefully controlled amounts of dopants to the silica tube. The dopants may be increased or decreased to raise or lower the refractive indexes of the host glass, which carefully defines the waveguide (fiber). This makes the DDS process uniquely suited for graded index waveguide fabrication. Graded index waveguides require a continuously varying, precisely controlled profile of the refractive index within the waveguide core.

A second class of methods leading from soot to chemical vapor deposition (CVD) produced a glassy layer rather than sootlike layer on the surface of a silica tube or the inside of the tube.

Figure 7–5 The Dope-Deposited Silica (DDS) Process (*Courtesy Corning Glass Works*)

The CVD process is by far the most widely utilized process in developing and manufacturing of waveguides. The process begins with an extremely high-quality tube of quartz, such as that shown in the figure 7–6. The glass tube's length is near a meter and its diameter is about 20 millimeters, depending on the fiber being processed and the manufacturing technique. The tube is placed in a special lathe (see figure 7–7) to create a preform which has the properties of the fiber. Heat is applied. A constant gas flow through the tube deposits boron-doped silica on the inner wall of the tube. The boron-doped silica has a lower refractive index than pure silica. This process provides a core material for the fiber. The borosilicate soot is uniformly deposited by the maintenance of a constant level of gas flow and a regulated flame along the tube. By repeating the process with varying amounts of dopant, graded or step index core regions may be built up. Germanium and phosphorus dopants are used to increase refractive indexes. Boron is used to lower refractive indexes.

Figure 7–6 High-Quality Quartz Tube (*Courtesy Western Electric*)

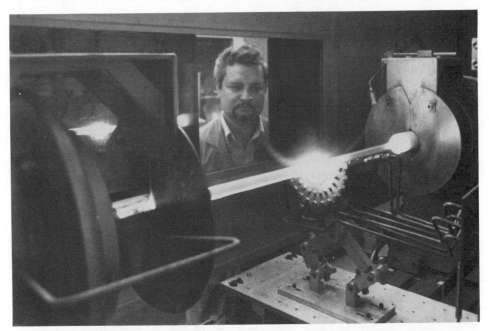

Figure 7–7 Preform Lathe (*Courtesy Western Electric*)

Figure 7–8 Quartz Tube after Deposition Process (*Courtesy Western Electric*)

Figure 7–8 is a quartz glass tube that has been through the deposition process. After a suitable number of layers of borosilicate has been deposited, the tube is collapsed to form the fiber preform (see figure 7–9). The preform is then placed into a specially designed lightguide drawing machine where the quartz preform is heated in an induction furnace or by CO_2 laser. In the machine, the preform is pulled into several kilometers of hair-thin lightguides. The lightguide drawing machine is a precision instrument. Figure 7–10, is a photograph of one of the drawing machines developed by Bell Laboratories and Western Electric.

It must be noted that the fabrication processes described here are simplified, because the scope of this book is introductory. The mechanics of fiber fabrication are critical and complex. Glass melting, flow rate, diffusion methods, geometry of containers, cooling, crucible loading and feeding procedures, atmosphere control, pulling/drawing, and winding are just a few of the exacting details involved in the fabrication process.

Figure 7–9 Collapsed Fiber Preform (*Courtesy Western Electric*)

FIBER TYPES AND SPECIFICATIONS

Fibers are constructed to fill the need of a particular requirement. The major need is in the area of bandwidth, that is, low, medium, high, or ultrahigh. In addition, a fiber may be constructed for its extremely low attenuation per kilometer. In some cases tensile strength is the most desired characteristic. Two operating classes of fibers are utilized, single mode and multimode. Single-mode fibers are fibers that carry one mode. Single-mode fibers must be designed to accommodate a specific wavelength, otherwise large attenuation will result. An attractive feature of the single-mode fiber is that it is not sensitive to microbending. These are losses induced by local lateral displacements of the fiber. Multimode fibers allow intermodally dispersed wavelengths. That is, the propagation of rays of the same wavelength follow different paths through the fiber, causing different arrival times.

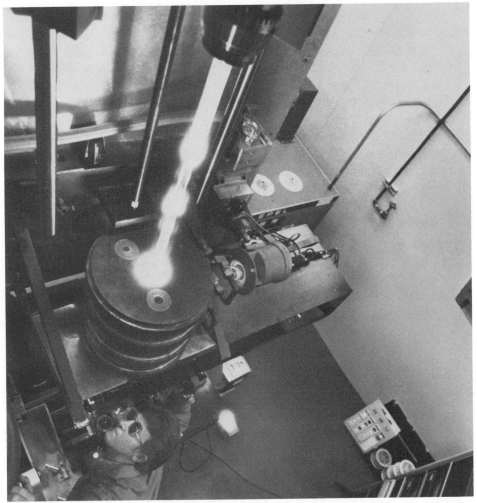

Figure 7–10 Lightguide (Fiber) Drawing Machine (*Courtesy Western Electric*)

There are two types of optical fibers, step index and graded index. Cross-sectional views along with the typical light-ray path of these fiber types are illustrated in figure 7–11. The step index fiber has an abrupt refractive index between the core and the cladding. The graded index fiber has a variation of refractive index between the core and the cladding. This minimizes dispersion and allows a greater bandwidth of information to be transmitted.

Fibers are constructed from plastic or glass or a combination of the two, e.g., glass core/glass cladding, plastic core/plastic cladding, or glass core/plastic

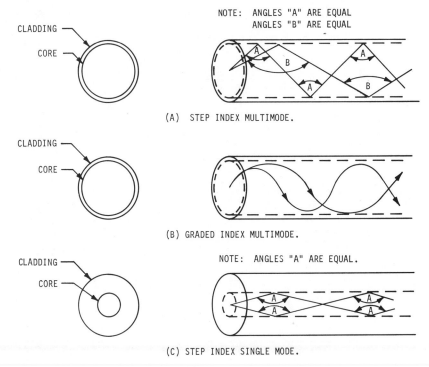

NOTE: ANGLES "A" ARE EQUAL
ANGLES "B" ARE EQUAL

(A) STEP INDEX MULTIMODE.

(B) GRADED INDEX MULTIMODE.

NOTE: ANGLES "A" ARE EQUAL.

(C) STEP INDEX SINGLE MODE.

Figure 7–11 Fiber Types (*Mike Anaya*)

cladding. The choice of a fiber type has to do with the quality, the parameters required, and economics. Can the system live with dispersion or must we use a better fiber at a higher price? In real life there are trade-offs in design. You cannot always receive lower cost without lower performance. A lower attenuation per kilometer may also mean a very high price tag.

In the next several paragraphs we shall take a look at several fiber types and their specifications. As you are comparing fiber types, consider what makes each more or less desirable than others.

Single-Mode Step Index Optical Fiber The single-mode step index designated is designed for use in very high bandwidth (500 MHz or more), single-fiber data systems, or in other applications that require single-mode propagation of light. The fiber consists of a doped silica core and a borosilicate cladding with a step-index profile. The cladding is coated with silica. A pair of jackets (one plastic) is extruded onto the fiber to provide mechanical and environmental protection (see figure 7–12).

INDEX OF REFRACTION PROFILE

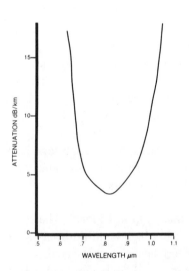

SAMPLE SPECTRAL ATTENUATION-SINGLE MODE STEP-INDEX OPTICAL FIBER

Figure 7–12 Single-Mode, Step-Index Optical Fiber (*Courtesy ITT Electro-Optical Products Division*)

Specifications	Nominal
Attenuation @ 0.63 μm and .85 μm	4–20 dB/km
Numerical aperture (NA)	0.1
Fiber core diameter	4.5 μm
Fiber outer diameter	80 μm
Jacket outer diameter	500 μm
Minimum bend radius	0.5 cm

Step-Index Multimode Fiber This glass step-index fiber (see figure 7–13) is designed for use in medium to high bandwidth single fiber data transmission systems. The fiber consists of a doped silica core and a borosilicate cladding with a step-index profile. The cladding is coated with silica. A pair of jackets (one plastic) is extruded onto the fiber to provide mechanical and environmental protection.

Specifications	Nominal		
Attenuation @ 0.85 μm	12 dB/km	— T101	As selected
	8 dB/km	— T102	As selected
	5 dB/km	— T103	As selected
Attenuation @ 1.06 μm	8 dB/km	— T101	As selected
	5 dB/km	— T102	As selected
	3 dB/km	— T103	As selected
Numerical aperture (NA)	0.25		
— 3dB intermodal dispersion	15 ns/km		
Core index of refraction	1.48		
Fiber core diameter	55 μm		
Fiber outer diameter	125 μm		
Jacket outer diameter	500 μm		
Tensile strength	5×10^5 psi		
Minimum bend radius	0.5 cm		

Graded-Index Multimode Optical Fiber This glass graded-index fiber is designed for use in high-bandwidth single-fiber data transmission systems. The fiber consists of a doped silica core with a graded index-of-refraction profile and a borosilicate cladding. The graded-index profile significantly reduces dispersion, thus permitting larger bandwidth signals to be transmitted. The cladding is coated with silica. A pair of jackets (one plastic) is extruded onto the fiber to provide mechanical and environmental protection. The fiber may be used singly or may be incorporated into multichannel fiber optic cables (see figure 7–14).

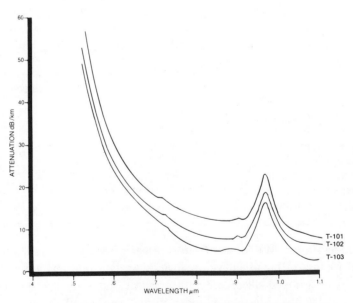

TYPICAL SPECTRAL ATTENUATION- STEP INDEX FIBER

Figure 7–13 Step-Index Multimode Fiber (*Courtesy ITT Electro-Optical Products Division*)

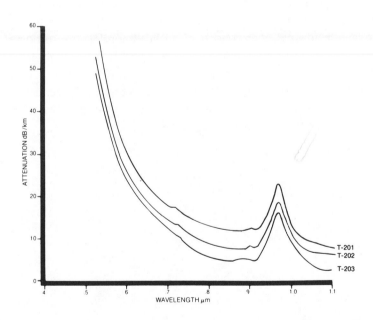

TYPICAL SPECTRAL ATTENUATION - GRADED INDEX FIBER

Figure 7–14 Graded-Index, Multimode Optical Fiber (*Courtesy ITT Electro-Optical Products Division*)

Specifications	Nominal
Attenuation @ 0.85 μm	12 dB/km — T201 As selected
	8 dB/km — T202 As selected
	5 dB/km — T203 As selected
@ 1.06 μm	8 dB/km — T201 As selected
	5 dB/km — T202 As selected
	3 dB/km — T203 As selected
Numerical aperture (NA)	0.25
−3 dB intermodal dispersion	3.5 ns/km
Core index of refraction	1.48
Fiber core diameter	55 μm
Fiber outer diameter	125 μm
Jacket outer diameter	500 μm
Tensile strength	5×10^5 psi
Minimum bend radius	0.5 cm

Wideband Graded-Index Multimode Optical Fiber   Figure 7–15 shows a wideband glass graded-index fiber that is designed for use in applications demanding the ultimate in channel capacity that present fiber-optic technology can provide. This ultrahigh-bandwidth fiber consists of a silica core doped to produce a graded refractive index profile, and a borosilicate cladding. The cladding is coated with silica. A pair of jackets (one plastic) provides mechanical and environmental protection. Bandwidths of 400 MHz can be achieved over 1 kilometer lengths of this fiber with narrow spectral-width sources. The fiber may be used singly or incorporated into multichannel fiber-optic cables.

Figure 7–15   Wideband Graded-Index, Multimode Optical Fiber (*Courtesy ITT Electro-Optical Products Division*)

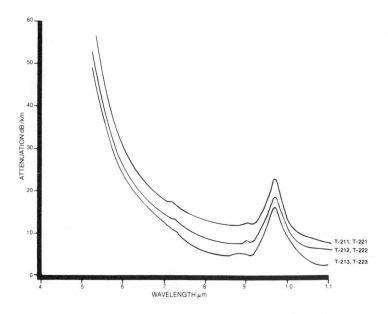

TYPICAL SPECTRAL ATTENUATION — WIDEBAND GRADED INDEX OPTICAL FIBER

Figure 7–15 *continued*

Specifications	Nominal		
Attenuation @ 0.85 μm	12 dB/km	— T211, T221	As selected
	8 dB/km	— T212, T222	As selected
	5 dB/km	— T213, T223	As selected
@ 1.06 μm	8 dB/km	— T211, T221	As selected
	5 dB/km	— T212, T222	As selected
	3 dB/km	— T213, T223	As selected
Numerical aperture (NA)	0.25		
—3 dB intermodal dispersion	2 ns/km	— T211, T212, T213	
	1 ns/km	— T221, T222, T223	
Core index of refraction	1.48		
Fiber core diameter	55 μm		
Fiber outer diameter	125 μm		
Jacket outer diameter	500 μm		
Tensile strength	5×10^5 psi		
Minimum bend radius	0.5 cm		

Large-core Plastic-clad Silica Multimode Optical Fiber   Large-core plastic-clad silica fiber is designed for use in economical, moderate distance, medium-bandwidth, single-fiber data transmission systems. The 200-micrometer core diameter offers greater source coupling efficiencies, which are useful when higher power levels are required at the receiver than could be provided by 125-micro-meter core plastic-clad silica fibers. The fiber consists of a homogeneous pure silica core surrounded by a plastic optical cladding and a plastic protective jacket. The core and plastic cladding form the optical waveguide, while the outer coating provides mechanical and environmental protection. The fiber may be used singly or incorporated into multichannel fiber-optic cables (see figure 7–16).

Specifications	Nominal	
Attenuation @ 0.79 μm	35 dB/km	— T321
	20 dB/km	— T322
	10 dB/km	— T323
Numerical aperture (NA)	0.3	
—3 dB intermodal dispersion	30 ns/km	
Core index of refraction	1.46	
Fiber core diameter	200 μm	
Jacket outer diameter	500 μm	
Tensile strength	5×10^5 psi	
Minimum bend radius	0.8 cm	

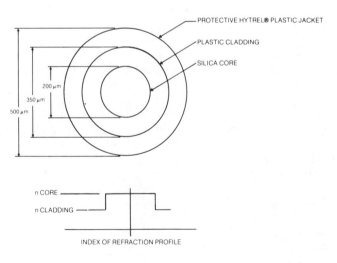

Figure 7–16   Large-Core, Plastic-Clad Silica Optical Fiber (*Courtesy ITT Electro-Optical Products Division*)

TYPICAL SPECTRAL ATTENUATION—LARGE CORE
PLASTIC CLAD SILICA OPTICAL FIBER

Figure 7–16 *continued*

Plastic-clad Silica Multimode Optical Fiber   Plastic-clad silica fiber is designed for use in economical, medium-bandwidth, medium-distance, single-fiber
data transmission systems. The fiber consists of a homogeneous pure silica core
surrounded by a plastic optical cladding and a plastic protective jacket. The core
and the plastic cladding form the optical waveguide, while the outer coating provides mechanical and environmental protection. The fiber may be used singly or
incorporated into multichannel fiber-optic cables (see figure 7–17).

Specifications	Nominal	
Attenuation @ 0.79 μm	35 dB/km	— T301
	20 dB/km	— T302
	10 dB/km	— T303
Numerical aperture (NA)	0.3	
−3 dB intermodal dispersion	30 ns/km	
Core index of refraction	1.46	
Fiber core diameter	125 μm	
Jacket outer diameter	500 μm	
Tensile strength	5×10^5 psi	
Minimum bend radius	0.5 cm	

TYPICAL SPECTRAL ATTENUATION - PLASTIC CLAD SILICA FIBER

Figure 7–17   Plastic-Clad Silica Optical Fiber (*Courtesy ITT Electro-Optical Products Division*)

CABLE TYPES AND SPECIFICATIONS

Cables of single or multiple fiber are manufactured usually for protection of the fiber. Fibers are encased within the cable using protective jackets (usually polyurethane). Strength members surround the fibers. Cables are constructed to feature light weight, high flexibility, resistance to kinks, strength, and resistance to crushing. Some tests imposed on cables in the development stage are impact, bend, twist, fatigue under load, high and low temperature, and even storage losses.

As in the choice of fibers, certain trade-offs are often necessary. However, the reasons for cable development are centered around strength and environmental conditions. These should not be traded off in the search for economics. The next several paragraphs contain information on cable types. The reader should compare each type and determine what the benefits are for using a particular cable.

Light-Duty Optical-Fiber Cable   The light-duty optical-fiber cable (see figure 7–18) is designed for use in single-fiber-per-channel transmission systems where a high-strength, rugged cable is not required. The standard cable contains up to 19 optical fibers. The optical fibers are encased in an extruded polyurethane jacket resulting in a small, flexible, lightweight cable. The cable is available with either plastic-clad silica, glass step-index, or glass graded-index fibers.

Specifications	Nominal
Number of fibers	Up to 19
Jacketed fiber diameter	500 μm
Cable diameter	2.5 mm
Weight	6.4 kg/km
Minimum bend radius	2.5 cm

Single-Fiber Strengthened Optical-Fiber Cable   (See figure 7–19.) The single-fiber strengthened optical-fiber cable is a small, strong, flexible, light weight cable design, suitable for many general purpose single-channel fiber optic links, such as between buildings, vehicles, or computers. The cable is suitable for installation in conduit, cable ducts, cabling trays, and for limited field use. The cable consists of a single optical fiber surrounded by six strength members and is encapsulated in an extruded polyurethane jacket. The cable is available with either plastic-clad silica, glass step-index, or glass graded-index fibers.

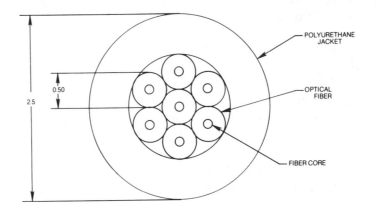

All dimensions in millimeters

*Other Configurations Available Upon Request.

OPTICAL SPECIFICATIONS (NOMINAL)*

CABLE TYPE[+] NUMBER[**]	OPTICAL FIBER DESCRIPTION	ATTENUATION (dB/km)	INTERMODAL DISPERSION (−3dB) (ns/km)	FIBER TYPE
T-461-XX	Glass Step	(.85 μm) 15		T-101
T-462-XX	Index Fiber	10	15	T-102
T-463-XX	(55 μm Core)	6		T-103
T-471-XX	Glass Graded	(.85 μm) 15		T-201
T-472-XX	Index Fiber	10	3.5	T-202
T-473-XX	(55 μm Core)	6		T-203
T-561-XX	Plastic Clad	(.79 μm) 35		T-301
T-562-XX	Silica Fiber	20	30	T-302
T-563-XX	(125 μm Core)	10		T-303
T-571-XX	Plastic Clad	(.79 μm) 35		T-321
T-572-XX	Silica Fiber	20	30	T-322
T-573-XX	(200 μm Core)	10		T-323

*Better performance available on a custom order basis.
**The last two type number digits indicate the number of active fibers in the cable.
+Formerly designated Type LD.

Figure 7–18 Light-Duty Optical-Fiber Cable (*Courtesy ITT Electro-Optical Products Division*)

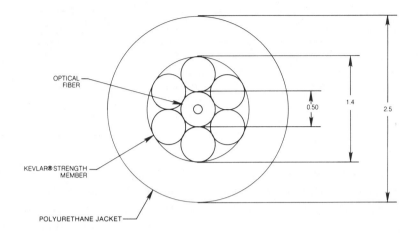

OPTICAL FIBER

KEVLAR® STRENGTH MEMBER

POLYURETHANE JACKET

All dimensions in millimeters

OPTICAL SPECIFICATIONS (NOMINAL)*

CABLE TYPE NUMBER[+]	OPTICAL FIBER DESCRIPTION	ATTENUATION (dB/km)	INTERMODAL DISPERSION (—3dB) (ns/km)	FIBER TYPE
		(.85 μm)		
T-421	Glass Step	15		T-101
T-422	Index Fiber	10	15	T-102
T-423	(55 μm Core)	6		T-103
		(.85 μm)		
T-431	Glass Graded	15		T-201
T-432	Index Fiber	10	3.5	T-202
T-433	(55 μm Core)	6		T-203
		(.79 μm)		
T-521	Plastic Clad	35		T-301
T-522	Silica Fiber	20	30	T-302
T-523	(125 μm Core)	10		T-303
		(.79 μm)		
T-531	Plastic Clad	35		T-321
T-532	Silica Fiber	20	30	T-322
T-533	(200 μm Core)	10		T-323

*Better performance available on a custom order basis.
+Formerly designated S1.

Figure 7–19 Single-Fiber Strengthened Optical-Fiber Cable (*Courtesy ITT Electro-Optical Products Division*)

Specifications	Nominal
Jacketed fiber diameter	500 μm
Cable diameter	2.5 mm
Weight	6 kg/km
Tensile strength	45 kgf
Minimum bend radius	2.5 cm

Two-Fiber Strengthened Optical Cable   The two-fiber strengthened optical cable (see figure 7–20) is designed for dual or duplex fiber-optic links within or between buildings, vehicles, or computers. It features small size, light weight, high strength, and flexibility. The cable is suitable for installation in conduit, cabling ducts, or cabling trays, and for limited outdoor use. The cable consists of two single-fiber cables, each composed of a central optical fiber. Each fiber is surrounded by six strength members and an extruded jacket. The cables are encapsulated in a black extruded polyurethane jacket. The cable is available with either plastic-clad silica, glass step-index, or glass graded-index fibers.

Specifications	Nominal
Jacketed fiber diameter	500 μm
Cable dimensions	4.4 mm $\times$ 2.5 mm
Weight	13 kg/km
Tensile strength	90 kgf
Minimum bend radius	3.5 cm

External-Strength-Member Heavy-Duty Optical-Fiber Cable   (See figure 7–21.) The external-strength-member heavy-duty optical-fiber cable is designed for use in single-fiber-per-channel transmission systems that require high cable strength and crush resistance. In addition, its high flexibility and kink resistance make it ideally suited for use in conduit, cable trays, and a variety of intravehicle applications. The cable contains up to 19 optical fibers in an extruded polyurethane jacket surrounded by helically laid strength members and an outer extruded polyurethane jacket. The design provides ease of strength member and optical fiber termination, high strength, and abrasion resistance. The cable is available with either plastic-clad silica, glass step-index, or glass graded-index fibers.

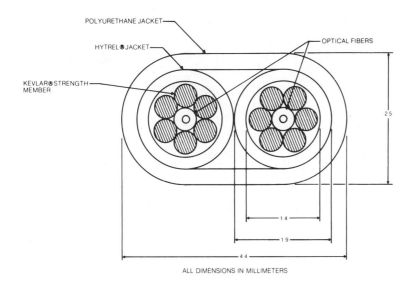

ALL DIMENSIONS IN MILLIMETERS

OPTICAL SPECIFICATIONS (NOMINAL)*

CABLE TYPE NUMBER	OPTICAL FIBER DESCRIPTION	ATTENUATION (dB/km)	INTERMODAL DISPERSION(—3dB) (ns/km)	FIBER TYPE
		(.85 μm)		
T-441	⌈ Glass Step	15		T-101
T-442	Index Fiber	10	15	T-102
T-443	⌊ (55 μm Core)	6		T-103
		(.85 μm)		
T-451	⌈ Glass Graded	15		T-201
T-452	Index Fiber	10	3.5	T-202
T-453	⌊ (55 μm Core)	6		T-203
		(.79 μm)		
T-541	⌈ Plastic Clad	35		T-301
T-542	Silica Fiber	20	30	T-302
T-543	⌊ (125 μm Core)	10		T-303
		(.79 μm)		
T-551	⌈ Plastic Clad	35	30	T-321
T-552	Silica Fiber	20		T-322
T-553	⌊ (200 μm Core)	10		T-323

*Better performance available on a custom order basis.

Figure 7–20 Two-Fiber Strengthened Optical Cable (*Courtesy ITT Electro-Optical Products Division*)

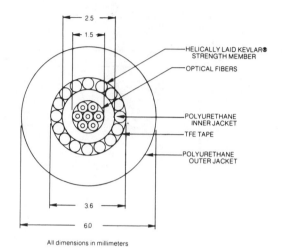

All dimensions in millimeters

*Other Configurations Available Upon Request.

OPTICAL SPECIFICATIONS (NOMINAL)*

CABLE TYPE[+] NUMBER[**]	OPTICAL FIBER DESCRIPTION	ATTENUATION (dB/km)	INTERMODAL DISPERSION (—3dB) (ns/km)	FIBER TYPE
		(.85 μm)		
T-401-XX	Glass Step	15		T-101
T-402-XX	Index Fiber	10	15	T-102
T-403-XX	(55 μm Core)	6		T-103
		(.85 μm)		
T-411-XX	Glass Graded	15		T-201
T-412-XX	Index Fiber	10	3.5	T-202
T-413-XX	(55 μm Core)	6		T-203
		(.79 μm)		
T-501-XX	Plastic Clad	35		T-301
T-502-XX	Silica Fiber	20	30	T-302
T-503-XX	(125 μm Core)	10		T-303
		(.79 μm)		
T-511-XX	Plastic Clad	35		T-321
T-512-XX	Silica Fiber	20	30	T-322
T-513-XX	(200 μm Core)	10		T-323

*Better performance available on a custom order basis.
**The last two type number digits indicate the number of active fibers in the cable.
+Formerly designated Type ESM

Figure 7–21   External Strength Member, Heavy-Duty Optical Fiber Cable (*Courtesy ITT Electro-Optical Products Division*)

Specifications	Nominal
Jacketed fiber diameter	500 μm
Cable diameter	6.0 mm
Weight	30 kg/km
Tensile strength	100 kgf
Minimum bend radius	5.0 cm

TRANSMISSION ANGLES

Transmission and power are dependent on many specifications and parameters. Perhaps the most important of these are the transmission angles. There are three angles involved in a step-index fiber. These are the angle of acceptance, the angle of refraction, and the angle of reflection.

If you recall from chapter 2, an incoming incidence ray, when reflected from a surface, reflects at the same angle as it strikes the surface. The angle of incidence equals the angle of reflection. The equal angles are also dependent upon the surface being smooth. The type of reflection is called *specular*. In figure 2–14, light refraction occurs at the surface that separates two different media such as air and water or air and optical fiber. The index of refraction is a ratio of the speed of light in a vacuum to the speed of light in a specific material. Each material has an index of refraction. Indexes for common materials are listed in tables of refraction in physics books. In the case of fibers, the manufacturer can provide this information. The refraction index is important when considering the angle of the light source at the fiber end.

Snell's Law

n_a sine $\angle$ a = n_r sine $\angle$ r

where n_a = index of refraction of air

        n_r = index of refraction of fiber

        $\angle$ a = angle of incidence

        $\angle$ r = angle of refraction

The physics of refraction is explained in terms of Snell's law (see figure 7–22). Rays intersecting a plane surface (fiber end) are refracted in the following relationship. The index of refraction of the air times the sine of the angle of incidence is equal to the index of refraction of the fiber times the sine of the angle of refraction. Now let's consider the angles related to optical fibers.

SNELL'S LAW

$$n_a \text{ SINE } \angle a = n_r \text{ SINE } \angle r$$

WHERE n_a = INDEX OF REFRACTION OF AIR.

n_r = INDEX OF REFRACTION OF FIBER.

Figure 7–22 Snell's Law (*Mike Anaya*)

Cone of Acceptance   The cone of acceptance is the area in front of a fiber that determines the angle of light waves that will be accepted into the fiber (see figure 7–23).

Angle of Acceptance   The angle of acceptance is an angle that represents the half-angle of the cone of acceptance (see figure 7–24). The mathematical sine of the angle of acceptance is called the *numerical aperture* (NA).

$$NA = \text{sine } \angle a$$

The reader will note that the angle a ($\angle a$) is usually designated $\angle \theta$ in engineering circles. It was felt, though, for explanation purposes, that we would use the first letter of the angle title. The numerical aperture (NA) can also be construed as the maximum acceptance angle which, in turn, defines the light-gathering power of the fiber. That is, the measure of the fibers ability to accept light waves and reflect them through the fiber.

Numerical aperture can be calculated by several means:

$$NA = \text{sine } \angle a$$
or $$NA = \sqrt{n1^2 - n2^2}$$
or $$NA = (n1^2 - n2^2)^{\frac{1}{2}}$$

where $n1$ = index of refraction of the core
$n2$ = index of refraction of the cladding
$\angle a$ = acceptance cone half angles

Angle of Refraction   The angle of refraction is the angle at which the rays are refracted from the cone of acceptance into the core (see figure 7–25).

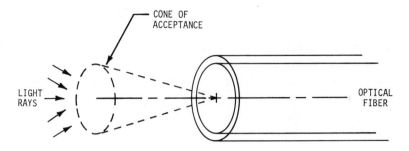

Figure 7–23 Cone of Acceptance (*Mike Anaya*)

$\angle$a = ANGLE OF ACCEPTANCE.

SINE $\angle$a = NUMERICAL APERTURE (NA).

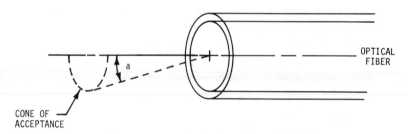

Figure 7–24 Angle of Acceptance (*Mike Anaya*)

$$\angle c2 = \frac{\angle a}{n1}$$

n1 = INDEX OF REFRACTION OF CORE.

n2 = INDEX OF REFRACTION OF CLADDING.

$\angle c1 = \angle c2$
$\angle r1 = \angle r2$

Figure 7–25 Transmission Angles (*Mike Anaya*)

Light waves are propagated from air at $\angle$a. These waves are refracted at the surface of the optic fiber into the core at $\angle$r. Therefore, the angle of incidence and the angle of refraction are related, using Snell's law. Calculations using Snell's law are as follows:

$$n_a \text{ sine } \angle a = n_r \text{ sine } \angle r$$

therefore $\text{sine } \angle a = \dfrac{n_r}{n_a} \text{ sine } \angle r$

Remember the numerical aperture NA = sine $\angle$a, which represents the maximum acceptance angle. Rays entering the fiber at a greater angle than $\angle$a are lost and will not reflect. Rays entering the fiber at a smaller angle than $\angle$a are totally reflected.

Angle of Reflection   Propagated waves at the angle of refraction ($\angle$r) strike the interface of the core and cladding. The angle of refraction at this point is actually an angle of incidence and the waves are reflected from the core/cladding interface at the same angle at which they strike the interface. In figure 7–25, the angles r1 and r2 are equal and the angles c1 and c2 are also equal. The angles c1 and c2 are called internal critical angles. Any mode that strikes the core/cladding interface at an angle greater than c will not be reflected but lost in the cladding. Modes that strike the interface at a smaller angle than c will be reflected through the fiber. The critical angles are also functions of the core and cladding indexes of refraction. The internal critical angle c2 is related to the fiber indexes of refraction in the following manner.

$$\angle c2 = \frac{\text{arcsine } \sqrt{n1^2 - n2^2}}{n1}$$

Graded-Index Fiber Angles   A graded-index fiber is used to reduce dispersion within the fibers (see figure 7–26). The numerical aperture (NA) is calculated in the same manner as with the step-index fiber. The graded-index core has an index of refraction that decreases with radial distance. The light waves travel faster near the cladding and slower near the core center. The phenomena allow the speed of all waves at all angles to be approximately the same. In actual practice there are three ray paths. Rays in the center of the fiber travel in high index of refraction (slow) material but travel the shortest path. Rays taking a sinusoidal path through the fiber cover a greater distance than the center rays, but travel in low-index-of-refraction (fast) material, part of the time. Finally, rays traveling in a helical path cover a greater distance, but travel in fast material (low index of refraction). This is an improvement over the step-index fiber in that larger angles in the step index take longer to reach the receiving end of the fiber.

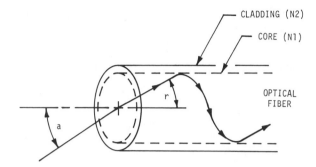

Figure 7–26 Graded-Index Fiber Ray Paths (*Mike Anaya*)

CAUSES OF COUPLING LOSS

Before we develop procedures defining alignment of optical fibers, it seems appropriate to discuss some of the causes of power loss in coupling. Power losses are rated in decibels (dB) and are additive regardless of their reason for existence. Some of these losses are as follows:

1. Heat is always a problem in dealing with any engineering effort. Optical fibers must be isolated with heat sinks tied to receptacles, mountings, connectors, and certain splices.
2. Fiber isolation, when in a cable or bundle, may cause some power loss. Sealing off adjacent light sources is necessary. Opaque fiber coverings and sheathing are often added to protect against light sources and the environment.
3. If cable ends are allowed to vibrate, the vibration may actually cause modulation of the light rays.
4. Sand or dust on the critical mirror-finished fiber ends may act as an abrasive. Pitted finish may reflect light rays in a diffuse rather than a specular manner.
5. Water or humidity at the connection may actually improve the coupling. Water has an index of refraction much nearer the fiber core than air.

These are just some of the reasons for power loss. The most prevalent of coupling power loss is due to misalignment of the fiber. There are four major forms of fiber misalignment. These are core irregularity, lateral misalignment, gap loss, and angular loss. A description of each is provided in the next several paragraphs.

Core Irregularity   During manufacture, the core of the fiber may be placed off center within the fiber. In extremely long fiber lengths there may be short stretches of core that are off center. In figure 7–27, cores are shown off center. The amount is exaggerated for understanding.

END
VIEWS

OPTICAL
FIBER

MISALIGNED
CORES

NOTE: EXAGGERATED CORE IRREGULARITY.

Figure 7–27 Core Irregularity (*Mike Anaya*)

Lateral Misalignment   The centerlines of cores may be misaligned. Axial misalignment, as this is often called, is the lateral displacement between centerlines of the cores in the two mating fibers (see figure 7–28). Power loss in decibels is plotted as an L/D ratio where L is the lateral displacement and D is the core diameter. Centerlines should be between 0.0001 and 0.0002 inch as specified, depending on fiber diameter.

Gap Loss   Gap loss is the separation of the two mirror finished fiber ends (see figure 7–29). Gap loss is expressed in decibels and is determined by the NA of the fiber. Lower value NA's have less gap loss than higher NA's. Gap loss is plotted as a G/D ratio against decibels where G is the gap and D is the core diameter. Typical allowable gaps are 0.0002 or 0.0003 or as specified.

Angular Loss   Angular loss is caused by cutting the fiber ends at an angle

Figure 7–28 Lateral Misalignment (*Mike Anaya*)

Figure 7–29 Gap Loss (*Mike Anaya*)

or by preparing the faces of the fibers so that they are not symmetric (see figure 7–30). Again, the power loss is expressed in decibels. Angular losses are plotted as angles in degrees against decibels, where the angle is the shape between the two mirror fiber end faces. Typical allowable angle is less than ¼ degree.

Figure 7–30 Angular Loss (*Mike Anaya*)

Loss in Transmitting from a Large Fiber to a Smaller One   When two fibers of different diameter are mated together a loss of optical energy will result. Calculation of this loss is made in the following manner:

$$dB = -10 \log \left(\frac{D_r}{D_t}\right)$$

where D_r = diameter of the receiving fiber

   D_t = diameter of the transmitting fiber

Losses in Transmission Due to Changes in Numerical Aperture (NA). When two fibers of different numerical aperture (NA) are joined together the loss of optical energy may be calculated in the following manner:

$$dB = -10 \log \left(\frac{NA_r}{NA_t}\right)$$

where NA_r = numerical aperture of receiving fiber

   NA_t = numerical aperture of transmitting fiber

Coupling Attenuation   Attenuation in coupling is established by two general criteria. These are intrinsic losses and extrinsic losses. Intrinsic losses are due to factors involved in the fiber and its manufacture. Intrinsic losses have nothing to do with the connector or its installation. Several producers of intrinsic loss are thus:

1. *Core Diameter Variations.* The core diameter variations are generally minute but can cause loss.
2. *Fresnel Reflection.* Fresnel loss is due to the difference in the index of refraction of the air and the fiber at the fiber end.
3. *Transmission Loss.* Losses due to the light waves lost in the cladding and not reflected from its walls.
4. *Absorption Loss.* Loss due to nontransparent core material.

Extrinsic losses are usually attributed to the connector hardware and man-made misalignment. Connectors and the details of misalignment and/or alignment are covered under separate headings.

ALIGNMENT PHILOSOPHY

Optimum alignment must be made each time a connection is made or broken and remade. This alignment must be easily obtainable without the need of special equipment, special handling, or tuning of transmitter and receiver. An extensive fiber alignment philosophy was presented at an Electronic Components Conference, Anaheim, California, on April 24–26, 1978, by Terry Bowen of AMP Inc., Harrisburg, Pennsylvania. This philosophy is generally recognized by the fiber-optic industry. This philosophy is provided in the next several paragraphs. Copyright © 1978 by AMP, Incorporated, Harrisburg, Pennsylvania. All international rights reserved. Reproduced with the permission of the copyright holder.

Point of Contact Analysis A primary alignment mechanism is formed when the common overlapping surface is in direct contact with the outer diameters of the fibers to be aligned (see figure 7–31). As seen in figure 7–32, optical fibers are generally round, formed of rigid material, and have outside diameters that vary in size as a result of manufacturing conditions. When brought into contact with a flat plane of rigid material, the axis of the fibers of different diameters will locate at different heights above the plane. See figure 7–33. There is no confinement in the horizontal direction so the fibers are free to roll about on the flat plane. Therefore, a flat plane does not provide a good alignment mechanism for optical fibers.

If two flat planes are made to intersect to form an included angle of 2ϕ, the resultant V-shape will prevent the round fibers from rolling (see figure 7–34). By making line contact at two points in the fiber periphery, the V-shape places the axis of any size round fiber in a common vertical plane which bisects the V. One drawback continues: Different-diameter fibers will have their axes placed at different heights above the bottom of the V-shape.

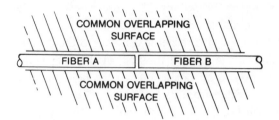

Figure 7–31 Primary Alignment System (*Courtesy AMP, Inc.*)

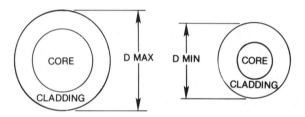

Figure 7–32 General Shape of Optical Fibers (*Courtesy AMP, Inc.*)

$$\Delta H = (H\ MAX)-(H\ MIN) = \frac{(D\ MAX)-(D\ MIN)}{2}$$

Figure 7–33 Core Height Misalignment (*Courtesy AMP, Inc.*)

Figure 7–34 The V-Groove Philosophy (*Courtesy AMP, Inc.*)

Formulas for height (ΔH) variations and maximum/minimum diameters (D) are listed below.

$$H(max) = \frac{D(max)}{2 \text{ sine } \theta}$$

$$H(min) = \frac{D(min)}{2 \text{ sine } \theta}$$

$$\Delta H = H(max) - H(min) = \frac{D(max) - D(min)}{2 \text{ sine } \theta}$$

It can be seen that the sharper the V-shape (the smaller θ becomes), the greater the difference in the height of the fiber axis. When optical fibers have identical diameters, a V-shape is an excellent alignment mechanism.

Three lines of contact are established on the round fiber if a third plane is placed over a shallow V-groove as shown in figure 7–35. In still another variation (see figure 7–36), three round rods are placed in contact and the round fiber is fitted into the interstitial space formed.

Figure 7–35 Three Lines of Contact (*Courtesy AMP, Inc.*)

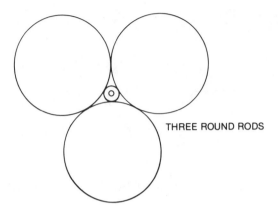

Figure 7–36 Round Rod Contact (*Courtesy AMP, Inc.*)

Four lines of contact can be made with opposing shallow V-shapes (see figure 7–37), or with four round alignment rods placed about the fiber. Attempts to form more than four lines of contact with circular-geometry tubes or sleeves produce, at best, a single-line contact effect with the fiber (see figure 7–38). Here, the fiber outer diameter must be smaller than the inner diameter of the tube, so it will tend to contact the tube at one point only.

Figure 7–37 Four Lines of Contact (*Courtesy AMP, Inc.*)

Figure 7–38 Circular Tube Single (*Courtesy AMP, Inc.*)

The common overlapping surfaces described are usually made of rigid material. A modified primary alignment mechanism can be made using resilient materials backed up with rigid materials (see figure 7–39). The inclusion of nonrigid or resilient materials provides a higher order of connector-alignment mechanism that successfully joins different-sized fibers.

Figure 7–39 Modified Primary Alignment Mechanism (*Courtesy AMP, Inc.*)

Ordinal Approach to Alignment Mechanism Figure 7–40 permits us to discuss the advantages and disadvantages of primary, modified primary, secondary, and tertiary alignment mechanisms.

Primary A rigid common overlap surface (1) straddles the fibers to be joined. On the plus side is simplicity, only one interface is made and that is directly with the fiber. The only tolerance involved is that associated with the fiber diameter and the surface(s) of (1). End preparation is simple using the score-and-break technique. The disadvantages come into play if this mechanism is used for connectors that must undergo repeated mating. There is a high risk of damaging or degrading the exposed fiber that is required to be interacting with the hard overlapping surface(s) of (1). For this reason the primary mechanism is best suited to permanent splicing where the fibers are inserted only once.

Modified primary The optical fibers are first covered by a resilient common overlay surface (1) then a rigid common overlay surface (1′). This concept is still a simple one; granted there are two interfaces, however the 1 to 1′ interface is not critical. The key here is that the 1′ to fiber A and B interface can absorb fiber diameter tolerances and 1′ is less likely to damage the fibers than is 1. Also, the resilient member provides a sealing capability. There is one drawback, the fiber surfaces are still exposed and may be subject to degradation.

Secondary The common rigid overlay surface (1) is still used, however, separate alignment components (2) are attached to fiber A and fiber B. These

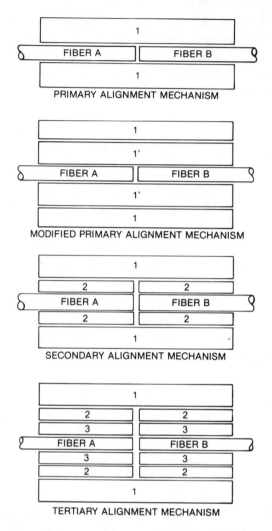

Figure 7–40 Ordinal Approach to Alignment Mechanisms (*Courtesy AMP, Inc.*)

may be rigid components or resilient components. In either case, the connector can be more rugged and can provide better protection for the fiber.

With all rigid components, two rigid-to-rigid interfaces must be made— each having tolerances to be concerned with. Attempts are made to keep these tolerances to a minimum, which naturally increases manufacturing costs.

A resilient component compensates for tolerance buildup just as in the modified primary alignment mechanisms, and is the subject of this paper.

Tertiary   This class adds another component layer (3), which brings with it a third interface with tolerances to consider. The advantages and disadvantages of tertiary alignment mechanisms are simply extensions of those of the secondary approach.

Resilient Alignment Mechanism   The resilient alignment mechanism for a single optical fiber is a secondary alignment mechanism with the rigid fiber being protected by a resilient elastometric ferrule. (See figure 7–41.)

Figure 7–41   Single-Fiber Resilient Alignment Mechanism (*Courtesy AMP, Inc.*)

Precise Hole Placement   The key to the success of the resilient alignment mechanism technique is the ultraprecise placement of this fiber centering hole— not in relationship to traditional alignment components, but actually in relationship to the true center that exists when the connector is in service. A proprietary manufacturing technique assures that the hole is on true center regardless of the size of the fiber or the initial shape of the ferrule. Ferrules are molded with the front end initially blank, and the small fiber-receiving hole precisely drilled during a secondary operation. (See figure 7–42.)

Basically, the blank ferrules are compressed into a circular drill fixture that forces the ferrules to assume their "in-service" shape. All irregularities in size and shape of the ferrule blank are transferred to variations in static loads at the periphery of the drill fixture.

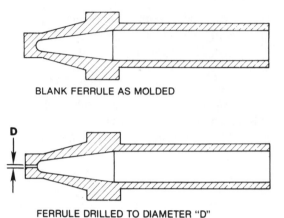

BLANK FERRULE AS MOLDED

FERRULE DRILLED TO DIAMETER "D"

Figure 7–42   Ultraprecise Ferrule Drilling (*Courtesy AMP, Inc.*)

The fiber-receiving hole is then drilled at the dead center of the blank. Ultraprecision required in the positioning of the drill fixture during the manufacturing setup is thereby introduced into the molded parts.

Precise Hole Manufacturing Process   Figures 7–43 and 7–44 follow the manufacturing process. The process for precise hole drilling in figure 7–43 is as follows:

  a. Inexpensive ferrule blanks may be slightly out of round.
  b. Ferrule blanks are compressed into drilling fixture which stimulates "in-service" condition of ferrule.
  c. Small hole is drilled, centered relative to drilling fixture; ultraprecise centering parts are now essentially identical.
  d. Free state after drilling; ferrule is usable now that ultraprecise hole positioning has been accomplished.

The process for insertion of the optical fiber is illustrated in figure 7–44.

  a. Fiber within acceptable tolerances is inserted with adhesive added.
  b. Ferrule is compressed to fit fiber size while adhesive sets.
  c. Faces of fiber and ferrule are polished to optical finish or the fiber is "broken" to form a mirror surface.
  d. The ferrule with fiber is placed into final alignment details.

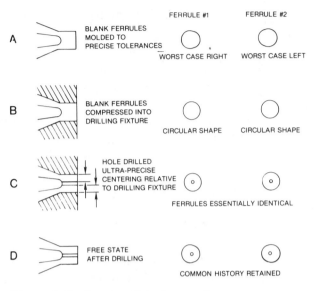

Figure 7–43 Precise Hole Shape (*Courtesy AMP, Inc.*)

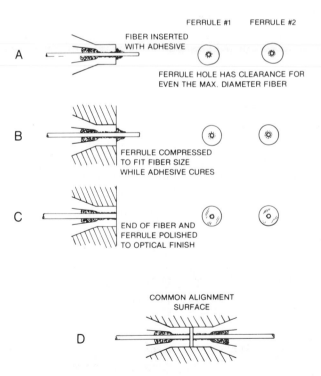

Figure 7–44 Insertion of Optical Fiber (*Courtesy AMP, Inc.*)

By producing the fiber-receiving hole of the ferrule to be slightly larger than the maximum fiber diameter, only one hole size is required. Compressive forces on the outer surface of the resilient ferrule are used to reduce the inner diameter producing a tight grip on even the smallest fiber size. A tight fit is achieved at the fiber-to-resilient ferrule interface and at the resilient ferrule-to-rigid overlap interface. The amount of compression achieved is determined by the particular size of the fiber involved. In all cases, the compression is sufficient to seal off the optical interface.

Benefits provided by the resilient alignment mechanism include:

1. The effect of inherent initial tolerance differences in fiber size are eliminated.
2. Ultraprecision "centered" connectors—by drilling the hole while the elastometric component is in compression—can be economically produced in volume.
3. A sealed interface protects the optical junction from hostile environments.
4. The rugged connector will survive many matings.
5. Resilient member cushions the fibers against damaging vibration and other mechanical problems.

ALIGNMENT MECHANICS

The mechanics of alignment are an exacting task in that perfect or at least optimum alignment is necessary to avoid loss. This philosophy must extend to the retaining of optimum alignment in every instance that a connection is made or remade. Furthermore, alignment must be of a method that can be accomplished in the field with a minimum of special equipment. Alignment should also be within the technical capabilities of available personnel who are already maintenance technicians.

Under the next several headings, we shall study several of the mechanical concepts used in the fiber industry today. Many of these were invented and patented by individuals and then were developed by major corporations for use in coupling devices.

Opposed Lens Alignment Concept   A pair of highly polished fiber ends are inserted into opposing cavities within a transparent medium. An objective lens and an immersion lens are provided by the opposed cavities in the transparent medium. The viscous fluid has an index of refraction that allows alignment of the opposing fibers. In figure 7–45, light rays are shown refracted from the viscous indices, through the transparent medium aligning opposing fibers. The inventor of the opposed lens alignment concept was Melvyn A. Holzman.

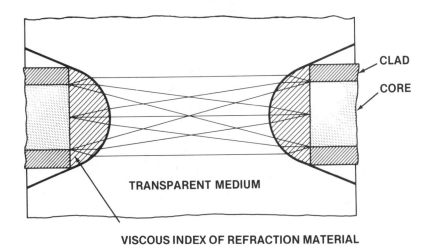

CLAD

CORE

TRANSPARENT MEDIUM

VISCOUS INDEX OF REFRACTION MATERIAL

Figure 7–45 Opposed-Lens Alignment Concept (*Courtesy ITT Cannon Electric*)

Transfer Molded Alignment Concept   Pairs of fiber end faces are highly polished to prevent Fresnel reflections, then insert molded on the axis of the resulting ferrule utilizing a fabricated and very precise tool (See figure 7–46). The two mating ferrules (plugs) insert within a biconical socket. Within the socket are cushions that are index matched to accommodate the plugs. This concept was defined by P. K. Runge and S. S. Cheng in the *Bell System Technical Journal,* July–August, 1978.

Multirod Alignment Concept   In figure 7–47, six precision rods are located around the inside perimeter of a guide sleeve. Three extremely precise ferrule rods provide a mounting for the fiber. An end view of the assembled ferrule and guide sleeve shows the fiber inserted at the exact geometric center within the three-rod ferrule triangle. All rods are calculated for tangency using needle bearings or cam rollers. Tolerances using this method can be held to millionths of an inch. The inventor of the multirod alignment concept was Carl R. Sandahl.

Resilient Self-Centering Alignment Concept   Two fibers of slightly different diameters may be aligned using the resilient self-centering alignment con-

Figure 7–46   The Transfer-Molded Alignment Concept (*Courtesy ITT Cannon Electric*)

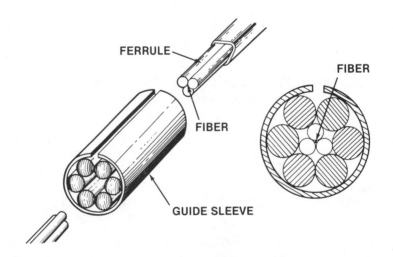

Figure 7–47   The Multirod Alignment Concept (*Courtesy ITT Cannon Electric*)

cept. In figure 7–48, the end view illustrates the fiber installed in a V-groove that establishes three planes of force on the fiber. Force is transferred equally to three arcs on the perimeters of the fibers. The material surrounding the fibers is resilient and will remember its original shape if the fibers are removed.

In the cross-sectional view A–A, the slightly different-diameter fibers are shown center aligned. Grinding, polishing, and bonding are accomplished using several standard procedures. The resilient self-centering alignment concept was invented by Robert M. Hawk.

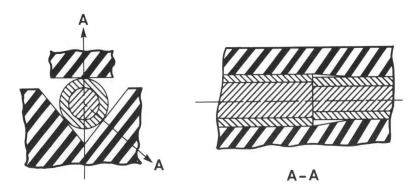

Figure 7–48 The Resilient Self-Centering Alignment Concept (*Courtesy ITT Cannon Electric*)

Double Eccentric Alignment   The double eccentric method of fiber alignment is one of the standard alignment procedures. The procedure deals with an eccentric motion similar to X–Y mechanisms. Adjustment is made in one direction or the other until fibers align. In figure 7–49 a pair of opposing ferrules is inserted into a guide cylinder and/or a male-female arrangement. The ferrules are utilized to adjust the fiber core to the axis of the outer cylinder surface. Adjustment is made under a microscope. When aligned, the ferrules are locked into place. The fiber ends are ground and polished. Epoxy is used to bond the fiber into the center sleeve.

A second method used with the double eccentric concept is to mate the fibers into a system, then adjust the eccentric until optimum power is realized at the receiver. The double eccentric concept was invented by Setsuo Sato.

Watch Jewel Concept   Readily available watch jewels are utilized to precisely align fibers as in the arrangement shown in figure 7–50. The jewels are produced at reasonable prices, with 3-micron concentricity inside or outside diameter. The alignment process involves measuring the diameter of the fiber with accuracy, then choosing the jewel to fit that is just larger in size than the fiber. The fiber is then epoxied in place and ground and polished to tolerance. The

Figure 7–49 The Double-Eccentric Alignment Concept (*Courtesy ITT Cannon Electric*)

Figure 7–50 The Watch Jewel Alignment Concept (*Courtesy ITT Cannon Electric*)

second fiber end is treated in the same manner. Finally, the two opposing jewel/ferrules are mated in a guide sleeve as pistons in a cylinder.

Three Sphere Concept Three spheres of exact diameter are placed in a pair of ferrule connectors (see figure 7–51). Each connector utilizes the three spheres in a plane of 120° increments with the spheres retained in a spherical race. In the center of the three spheres, a defined space for the fiber end is positioned. When the two pairs of ferrules are brought together, the two sets of spheres nest with respect to each other at 60° increments, precisely aligning the optical fibers. Alignment accuracy is influenced only by the variation in sphere diameter. Spheres are available with ten-millionths of an inch tolerance.

Figure 7–51 The Three-Sphere Alignment Concept (*Courtesy ITT Cannon Electric*)

FIBER END PREPARATION

General Before placing a fiber into a connector, it must be properly prepared. The jacket of the fiber must be removed, the fiber is then broken and the fiber end is ground and polished. Final procedures also involve cleaning and coating the fiber with a suitable refractive-index-matching compound. Procedures below are utilized for both step-index and graded-index glass fibers.

Jacket Removal The initial step in fiber end preparation is the removal of the fiber jacket. In figure 7–52, the fiber is taped to a flat surface and is placed under tension by pulling. The plastic is carefully stripped away using a razorlike knife (X-Acto) or a razor. The knife is placed on a shallow angle ensuring that there is little or no abrasion of the fiber. It is also possible to remove the jacket using a wire stripper of the correct size. Wire strippers are used only as an alternative. Once the fiber jacket is removed, the fiber is cleaned (see figure 7–53). Soft tissue soaked in isopropyl alcohol or Freon is usually the agent for cleaning. The bare fiber (see figure 7–54) is now ready for end cutting and finishing.

Figure 7–52   Removing the Fiber Jacket (*Courtesy ITT Electro-Optical Products Division*)

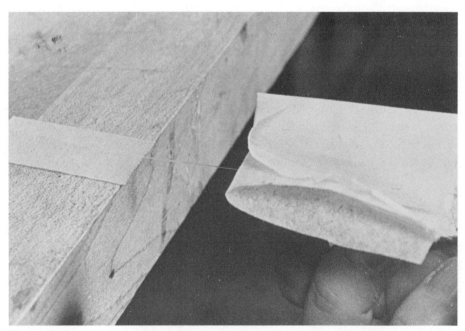

Figure 7–53   Cleaning the Fiber (*Courtesy ITT Electro-Optical Products Division*)

Figure 7–54 A Bare Fiber Ready for Finishing (Enlarged) (*Courtesy ITT Electro-Optical Division*)

Scribing and Breaking the Fiber   Before placing a fiber into a connector or splice it must be cut to size. This is accomplished using one of several techniques. Most are simple and can be easily learned.

Quartz and glass fibers are cut by scribing. The first involves scribing a line around the fiber with a fine file. Then by application of a constant pressure pull on both ends of the fiber and it will snap in two. This method is not very efficient when dealing with small-diameter fibers. It will, however, provide large-diameter fibers with a clean break where the line was scribed. In place of the fine file, a sapphire or a diamond cutter may be used for smaller fibers (see figure 7–55). A steady tension on the fiber will snap them. This tension is not large (around 0.25 pound-feet). The force, though hand-applied, will snap the fiber

Figure 7–55 Scribing the Fiber (*Courtesy ITT Electro-Optical Products Division*)

end smooth (see figure 7–56). There are other methods for cutting. These are commercial setups with special tooling to be purchased. Most follow a scribing philosophy of some sort. Plastic fibers may be cut with a razor blade.

Figure 7–56 Fiber End after Breaking at Scribe Mark (Enlarged) (*Courtesy ITT Electro-Optical Products Division*)

Inserting and Sealing Fiber into Ferrule If the fiber is to be placed into a ferrule for further installation in a connector, the end must be lapped and polished. The fiber end is measured using a micrometer (see figure 7–57) to ensure that the fiber and the ferrule are matched. In this procedure, the fiber end is broken and inserted into the end of the ferrule. Figure 7–58 illustrates the hand alignment necessary to place the fiber into the ferrule. Figure 7–59 shows the fiber in the ferrule with part of the fiber end protruding from the ferrule tip. A closeup look at the inserted fiber is provided in figure 7–60. Epoxy application to the fiber and the ferrule is now to be accomplished. In figure 7–61, a cross-sectional line drawing of the inserted fiber in the ferrule is presented. The plastic-coated fiber is surrounded in a bed of epoxy resin with the bare end of the fiber aligned in a watch jewel. Figure 7–61 presents the assembly after epoxy application. The next several illustrations describe the process.

Epoxy is now to be applied to the ferrule. Epoxy should have a low viscosity, good adhesion to the stainless steel ferrule and the glass fiber, and reasonable life. The epoxy should also have a short cure time, but not so short that it will prevent the ferrule from being filled prior to hardening. Epoxy recommended by the vendor should be used whenever possible, for they have taken the research time to choose the most compatible type.

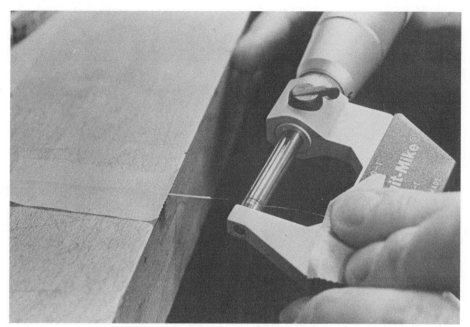

Figure 7–57 Measuring the Fiber Diameter (*Courtesy ITT Electro-Optical Products Division*)

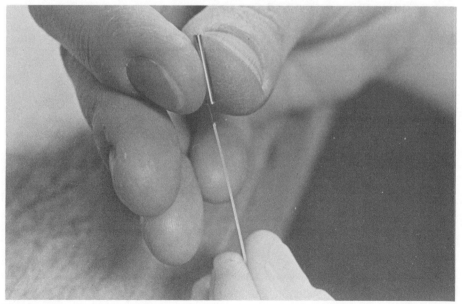

Figure 7–58 Hand Aligning of Fiber with Ferrule (*Courtesy ITT Electro-Optical Products Division*)

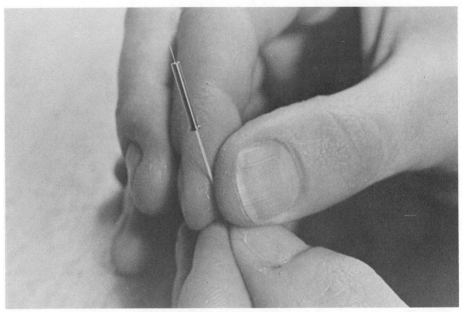

Figure 7–59 Fiber in Ferrule (*Courtesy ITT Electro-Optical Products Division*)

Figure 7–60 Closeup of Fiber in Ferrule (Enlarged) (*Courtesy ITT Electro-Optical Products Division*)

Figure 7–61 Cross Section of Fiber in a Stainless Steel Ferrule (*Courtesy ITT Electro-Optical Products Division*)

The fiber is attached to a termination station such as that drawn in figure 7–62. Epoxy is applied to the front and then the back of the ferrule as shown in figure 7–63. The ferrule should be heated to approximately 80°C in the heating fixture attached to the fiber-termination station. The epoxy should wick (flow down thin cracks) completely into the ferrule from the back. If it does not, a syringe may be used to draw epoxy the rest of the way into the ferrule (see figure 7–64). To speed the curing of the epoxy the assembly is kept in the termination station and heated to 80°C for about an hour or until the epoxy is cured. The ferrule, as it appears in figure 7–65, is ready for lapping and polishing.

Figure 7–62 Fabric Termination Station (Brass) (*Courtesy ITT Electro-Optical Products Division*)

Figure 7–63 Application of Epoxy to the Ferrule (*Courtesy ITT Electro-Optical Products Division*)

Figure 7–64 Drawing Epoxy into Ferrule (*Courtesy ITT Electro-Optical Products Division*)

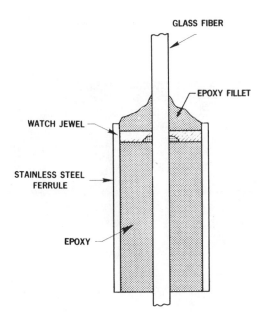

GLASS FIBER

EPOXY FILLET

WATCH JEWEL

STAINLESS STEEL
FERRULE

EPOXY

Figure 7–65 Fiber and Ferrule just after Epoxy Is Applied (*Courtesy ITT Electro-Optical Products Division*)

Lapping and Polishing Fiber Ends   Fiber ends must be ground, then polished to a mirror finish. Grinding and polishing takes place in several stages from rough to finish. The fiber is placed in a pin vise or some other type of holder so as to have freedom of movement. Grinding usually takes place in a mixture of alcohol and water.

Several grades of silicon carbide finishing paper are placed on a firm, flat surface. The lapping and polishing tool is held firmly in the hand and rotated in circular, elliptical, and figure-eight patterns on the surface of the finishing paper (see figure 7–66). The process is started on the rough paper and moves toward the finer grit. Before moving to a finer-grit paper, the surface of the fiber end is gently brushed off and cleaned in alcohol to remove loose silica chips. Throughout the process the fiber end is inspected and lapping is slowed when the ferrule shows evidence of having been ground flat. The front face of the jewel may be slightly recessed with respect to the ferrule at this point in the procedure. Lapping may be continued with light pressure to bring the jewel and ferrule planes just into or slightly short of coincidence (see figure 7–67), with a slight film of epoxy remaining over the ferrule endface. This procedure generally requires 1 to 3 minutes. Plastic fibers do not require as much finish work as glass or quartz.

When polishing is finished, a coating of a compound that has near the refractive index of the glass is applied to the fiber end. This compound will help to decrease attenuation losses.

Figure 7–66   Lapping the Fiber End (*Courtesy ITT Electro-Optical Products Division*)

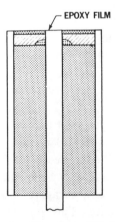

Figure 7–67   Fiber End after Lapping (*Courtesy ITT Electro-Optical Products Division*)

Before proceeding with polishing, the ferrule and lapping tool are thoroughly washed to remove residue grit remaining from lapping. A polishing paste which has suspended is used as the polishing compound. This paste is applied to moist adhesive-backed paper which has been mounted on a firm surface (see figure 7–68).

Figure 7–68   Polishing the Fiber End (*Courtesy ITT Electro-Optical Products Division*)

Depending on the smoothness of the lapped surface, a well-polished surface can be obtained in as little as 1 minute using moderate pressure. The fiber surface will become very slightly concave and recessed with respect to the face of the jewel (see figure 7–69). This characteristic is highly desirable for protection of the fiber during repeated connector matings. A photograph of the polished fiber and ferrule is provided in figure 7–70.

Inspection Criteria   After polishing and cleaning, the terminated ferrule end should be inspected. Acceptable terminations should look somewhat as the two photographs on the left side of the figure 7–71. Note the mirrorlike finish. The fiber should have no cracks in its core. The surface of the fiber should be flush with the face of the jewel or slightly concave. Under no circumstances can the fiber be convex. The photographs in the right side of figure 7–71 illustrate unacceptable terminations. As shown, these fibers are under 400× magnification.

Figure 7–69 Cross Section of Polished Fiber and Ferrule (*Courtesy ITT Electro-Optical Products Division*)

Figure 7–70 Lapped and Polished Fiber End in Ferrule (Enlarged) (*Courtesy ITT Electro-Optical Products Division*)

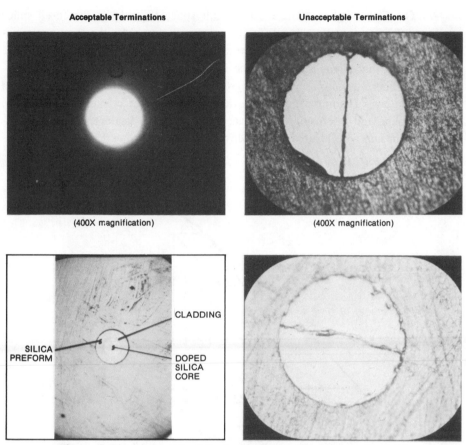

Figure 7–71 Inspection Criteria (*Courtesy ITT Cannon Electric*)

SPLICING

There are dozens of different splices and splicing techniques used in the fiber-optic industry. All splices have several aspects in common. They all must have the attribute of low power loss. Attenuation must be held to as near 0.1 decibel as possible. Splices must be quick to install, should be lightweight, extremely strong, and small.

Probably the most critical requirement of splicing is alignment of the fiber. Alignment philosophy and mechanics were covered in detail in previous paragraphs. Although fiber-optic manufacturers provide a large variety of splices, most of these utilize several basic techniques.

The Fused Splice The most difficult splice to make is the fused splice (see figure 7–72). It is also the splicing technique that provides the lowest attenuation. Economics and efficiency are a trade-off. Two fiber ends are placed in a tooling fixture that aligns the ends to be fused. Fiber ends are polished and prepared. The fibers usually are set in V grooves in the fixture and are clamped in resilient vise jaws. Heating elements are then brought to the connecting fiber ends. Heat is applied and the fusing takes place. Several fusion heaters used are microtorches, nichrome wire heaters, and electric arc fusers.

Figure 7–72 The Fused Splice (*Mike Anaya*)

The Siecor Model 65 Fusion Splicer This method employs the recognized method of permanently fusing two butted optical fibers together by heating with a tiny, well-regulated electric arc (see figure 7–73).

The field-usable splicer consists of fiber-holding and alignment devices, a viewing microscope, arc positioning and intensity controls, a built-in arc power supply (which requires only a 12-volt dc source), and a fiber cutting tool. Typical splice losses between telecommunications-grade optical fibers achieved with this device are from 0.2 to 0.4 decibels.

Field experience has shown that the Model 65 requires no extensive operator training or experience for the production of low-loss splices. After the

Figure 7–73 The Siecor ® Model 65 Fusion Splicer (*Courtesy Siecor Optical Cables Inc.*)

fibers are cleaned and cut they are clamped in the holders and positioned with the aid of the microscope. The arc is used briefly to burn off the fiber coating and then to slightly round the fibers' ends. With a higher arc power, the fibers are fused together. Monitoring splice loss during the actual operation is desirable but not necessary. After splicing, the joint is recoated and placed in a protective sleeve before going into the splice closure.

The Precision Tube Splice   A second method of splicing is the precision tube method (see figure 7–74). This method utilizes a precision tube that has been machined to a close tolerance and will accept the fiber diameter. Fiber ends are polished and prepared. A splicing compound may be injected into the splice. The splicing compound has the same refractive index as the fiber to be spliced. The fiber ends are inserted into the splice. A precision guide may be used to accommodate the insertion. The fiber ends are butted together within the splice and the outer jacket of the fiber is crimped. Care must be taken during crimping not to disturb the cladding and core interface. Crimping tools suggested by the fiber manufacturer are an absolute must. In some cases, the fiber splice is surrounded by a metal case or sleeve. The sleeve is crimped so as not to disturb the critical makeup of the fiber. Thermal shrink tubing is an acceptable alternative to crimping.

SPLICE

OPTICAL
FIBER

OPTICAL
FIBER

Figure 7–74   The Precision Tube Splice (*Mike Anaya*)

The Loose Tube Splice   Still another splicing method called a loose tube splice, uses a rectangular tube for splicing (see figure 7–75). The tube has a square hole that is curved slightly within the splice. Fiber ends are polished and prepared. A splicing epoxy is injected into the splice. The epoxy has the same refractive index as the fiber to be spliced. The fiber ends are inserted into the splice and must be slightly bent to insert. The fiber ends are butted together. Rotation of the square tubes guides the fiber ends together in the V groove in one corner of the splice. Once the alignment takes place, the epoxy is allowed to set (cure).

The Double Metal Plate Splice   A splicing method that is widely used in some form or another is the double metal plate splice (see figure 7–76). Also called the grooved substrate splice, this technique involves two machined metal

Figure 7–75 The Loose Tube Splice (*Mike Anaya*)

plates. The plates have a V-groove down their center which has been geometrically matched to the size of the fiber to be spliced. A copper or silicon substrate may also be coated on the metal backing and etched to ensure the correct V-groove to match the fiber diameter. Fiber ends are polished and prepared. A splicing epoxy is placed in the V-grooves on each metal plate and on the fiber ends. The epoxy must be refractive index matched to the fiber. The fiber ends are placed on one machined plate and the ends are butted together. The second machined plate is placed over the V-groove, aligned, and installed on the first plate. Suitable methods of fastening the two metal plates together are applicable.

Figure 7–76 The Double-Metal-Plate Splice (*Mike Anaya*)

The V-Shaped Splice   The Siecor Model 60 Mechanical Splicer is a small, portable device (see figure 7–77) which can connect the fibers of two or more optical cables in a simple, reliable manner with no need for touching or handling the fiber. It consists of two mounted swiveling clamp assemblies to hold the buffer jackets and fibers, a sliding tool to scribe and break the fibers to a precise, preset length, a cam-operated fixture for holding and positioning a V-groove splice part, and means for holding a splice case internal mounting plate. All components are mounted on a base plate. Also incorporated in the splicer is a fixture to hold the splice closure's internal mounting plate, which greatly simplifies the procedure of making individual fiber splices with multifiber cables. Figure 7–78 illustrates the splicing sequence.

The Siecor technique is based upon the principle of centering the fibers to be joined on the bottom of a V-shaped alignment part after breaking the fibers normal to their axes and at the exact required length. A magnetic support holds the V-groove alignment part and is cam-actuated to lift it to the exact alignment position. By raising this V-shaped splice part upward, the fiber ends, as a result of fiber stiffness, are pressed together and are aligned themselves on the bottom of the V-groove. A droplet of transparent, two-component epoxy holds the fiber ends in position, acting as an index-matching substance at the same time. After

Figure 7–77   The Siecor ® Model 60 Mechanical Splicer (*Courtesy Siecor Optical Cables, Inc.*)

BUFFER
JACKET

FIBER

SPLICE
PART

HOLD
DOWN

PROTECTIVE
TUBING

EPOXY
CEMENT

KEEPER

Figure 7–78 Splice Sequence Using the Siecor Model 60 Mechanical Splicer (*Courtesy Siecor Optical Cables, Inc.*)

the epoxy has cured, the buffer jackets of the fibers spliced together are clamped by the serrated ends of the splice parts.

There are, as you may have deduced, many other splicing techniques. Roller-bearing rods externally surrounding a fiber splice and held together with shrink tubing are used. Precision-molded plastic, machined metal, or etched silicon have been extensively involved with fiber bundles. Proprietary splices will always be a significant part of the market. Until industry can standardize splicing and connecting fiber ends, the intricacies of the techniques involved will be an economical marketing entity.

CONNECTORS AND THEIR INSTALLATION

Connector Basics Connectors are required to join fibers at transmitters (light sources), receivers (detectors), and bulkheads. That is, the connector may be used to join fiber to fiber or device to fiber. Whichever the situation, the connector must fulfill the requirements of alignment so as to avoid the problem of attenuation. The subject of alignment has been brought up many times in this chapter and necessarily so. However, that is just one of the demands placed upon connectors. The connector must provide termination for jackets, strength members, and buffering materials at the fiber's end. It also must ensure that loads applied are not transferred from one fiber to the next or from a fiber to a cable or vice versa. Other uses of the connector are protection of the fiber during installation and for fiber-end preparation.

In separable connectors (connectors that have repeated matings), it is extremely important that the two fiber ends do not meet or touch. The reason being is that repeated matings due to shock or vibration will chip or scratch the polished fiber end and thereby degrade the optical efficiency and the performance of the fiber, therefore the system. The gap, however, cannot be large due to gap loss attenuation. The standard gap usually does not exceed 10 percent of the core diameter. An example is a 5-micron (0.0002-inches) gap for a 50-micron (0.002-inch) core.

Connector Types   Connectors are designed to provide connection for single fiber, multichannel, bundle fiber, etc. Figure 7–79 is an illustration of a variety of fiber-optic connectors. Two optical cable terminations and a threaded sleeve illustrate how an in-line connection can be made that gives insertion loss less than 1 decibel. (See figure 7–80.) The sleeve's lining compensates for allowable variations in termination dimensions and tends to self-center the assembly. Such a connection is also suitable for a feed-through, bulkhead application. Receptacles containing light-emitting diode (LED) sources and PIN photodetector diodes provide a means of mechanically and optically connecting a terminated optical cable to a transmitter and receiver. (See Figure 7–81.) The mounting of the receptacles is compatible with normal PC-board practice.

Design Considerations   Alignment considerations are the basic starting points in the design of any fiber-optic connector. For the mating of opposing fibers, each degree of freedom of alignment is responsible for a loss, which is additive. Optical isolation between adjacent ferrules must be provided by sealing off the light from each other.

Heat-dissipation techniques must be designed into receptacles for device mounting. Electrical isolation must also be considered between adjacent channels for the electro-optical devices. Cable strength members must be terminated within the connector such that any axial load on the cable is resisted by the connector hardware and its mounting, and not transferred to the fiber. This precludes the possibility of varying alignment and subsequent coupling efficiency change.

Figure 7–79 A Variety of Fiber-Optic Connectors (*Courtesy ITT Cannon Electric*)

Figure 7–80 In-Line Connectors (*Courtesy Siecor Optical Cables, Inc.*)

Figure 7–81 Source or Detector Connector (*Courtesy Siecor Optical Cables, Inc.*)

The effects of environment are different on an optical interconnection. For example, humidity may improve coupling efficiency since the refractive index of water is better matched to the fiber's index than that of the air between mated fibers. Sand and dust are a more severe environmental condition than for electrical interconnections. The sand used in MIL–STD–202 (test methods for electronic and electrical component parts) testing is the same silica as the fiber and, therefore, erodes the fiber's polished endface. Vibration tends to modulate the optical pulses if the ferrules vibrate independently of each other.

Connector Installation Procedures   The process of connecting as a norm follows a rather exacting procedure. The fiber-optic connector is a much more critical and refined item of hardware than the standard electrical connector. Procedures for installing a connector on a fiber are usually developed for the peculiar fiber and by the special features of the connector itself. However, most connectors do follow general procedures. These are as follows:

1. The fiber jacket and buffering material are removed from the end of the fiber or cable of fibers.
2. The exposed fiber is broken for the precise length required. Fixtures are available for producing controlled, consistent breaks.
3. The fiber must be ground and polished.
4. Cleaning the fiber is a necessity. Usually this takes place between any step in the entire procedure.
5. The fiber must be aligned precisely in the connection device. This critical procedure is usually accomplished using special fixtures. However, in the field assembly may simply require a connection that is manufactured to fulfil the philosophy of alignment using good mechanics. Low attenuation due to precisely constructed connectors is the goal.

Figure 7–82 is an installation procedure incorporating the general procedures listed above. The figure is the procedures for installing an optimate single-position fiber-optic connector. Figure 7–83 is a second installation using general procedures. This second procedure is for an optimate multiple-position fiber-optic connector. All procedures for connector installation are similar. Industry is attempting to standardize the process.

Figure 7–82 Optimate Single-Position Fiber-Optic Connector Installation Procedures (Sheet 1 of 2) (*Courtesy AMP Inc.*) ▶

TYPICAL CABLE CONNECTOR ASSEMBLY

CRIMPING RING P/N 530526-1

RETAINING CAP P/N 861403-7

JACKET

OPTIC FIBER

POLISHING BUSHING P/N 530522-1

FERRULE (COLOR CODED)

FERRULE SHOULDER

CABLE SIZE (O.D.)		CABLE CONNECTOR KIT P/N	FERRULE	
JACKET	OPTIC FIBER		COLOR	P/N
.063	.017	530530-6	BLUE	530954-6
.063	.020	530530-7	VIOLET	530954-7
.075	.017	1-530530-0	BLACK	1-530954-0
.075	—	530530-3	ORANGE*	530954-3
.090	.020	530530-4	YELLOW	530954-4
.090	.045	530530-2	RED	530954-2
.095	.024	530530-5	GREEN	530954-5
.100	.017	530530-9	NATURAL	530954-9
.120	.045	530530-1	BROWN	530954-1
.120	.063	530530-8	GRAY	530954-8
.125	.017	1-530530-1	BROWN	1-530954-1

* This ferrule is designed for unstripped .075 In. dia. cable (jacket and optic fiber).

Figure 1

INPUT/OUTPUT BUSHINGS

STANDARD MOUNT

PRESS-FIT MOUNT

.468 (TYP)

.140 DIA

A DIA   B DIA

C   D

DIMENSIONS				BUSHING P/N	ACTIVE DEVICE PACKAGE TYPE
A	B	C	D		
.430	.330	.200	.295	530524-1	TO-5
.278	.191	.110	.295	530563-1	TO-18, TO-46
.234	.188	.684	.130	530565-1	MOLDED LENS
.430	.330	.080	.295	530525-1	TO-99
.278	.191	.326	.125	530564-1	TO-18, TO-46
PRESS-FIT				227015-1	

Figure 2

† Trademark of Loctite Corp.

★ Trademark of AMP Incorporated

1. INTRODUCTION

The AMP Optimate Single Position Fiber Optic Connectors consist of two parts: a cable connector assembly, as shown in Figure 1, and a connector bushing, as shown in Figure 2 or 3. Each cable connector assembly consists of a crimping ring, a retaining cap, a color-coded ferrule, and a polishing bushing. The cable connector assembly components are available individually or in kit form.

There are two types of connector bushing: input/output and splice. The input/output bushings provide a means of interfacing the fiber optic cable to the source or detector devices. The splice bushings allow a way of making a disconnectable joint between cables.

2. CABLE TERMINATION

(a) Slide crimp ring and retaining cap onto cable and then carefully strip 1/2" of jacket from end of cable. Do not damage fiber. Strip glass bundles with hot wire strippers. Remove all strength members.

(b) Clean ferrules ultrasonically in alcohol to remove contaminants and lubricants. Clean glass ultrasonically in alcohol, be sure no loose pieces of stranding remain in bundle.

(c) For glass bundles: Apply 3 drops of glue to fibers. Gently fan out strands and thoroughly mix glue into the bundle. Apply one drop of glue to the jacket. Insert cable into ferrule as far as it will go. See Figure 4.

NOTE: *Glues recommended for glass fibers are LOCTITE† 414, LOCTITE 495, or equivalent.*

For plastic fibers: Apply one drop of solventless epoxy glue (2 part, 100% solids) to the fiber and one drop to the jacket. Insert cable into ferrule as far as it will go. See Figure 4.

CAUTION: USE ONLY SOLVENTLESS, 2 PART, 100% SOLIDS, EPOXY GLUE. Any other type will damage fiber cladding.

BULKHEAD MOUNT

TYPICAL SPLICE BUSHING

.312 DIA O.D. (HOLE IN PANEL MUST BE SLIGHTLY LARGER)

FREE-HANGING

.750

.120 DIA

BUSHING P/N	TYPE
530570-1	BULKHEAD MOUNT
530523-1	FREE-HANGING BUTT

Figure 3

(d) Slide retaining cap over ferrule, followed by crimp ring. Position crimp ring about 1/16" behind cap so that cap can rotate freely when installed. See Figure 4. Crimp with tool P/N 90364-1, as shown in Figure 5.

(e) Carefully trim fiber so that it protrudes 1/16" (approx.) from tip of ferrule, as shown in Figure 4. Do not break fiber.

(f) Screw polishing bushing into retaining cap until ferrule is firmly bottomed. Tip of ferrule will protrude .020" (approx.) from bushing. See Figure 6.

(g) Place 5 grades of silicon carbide finishing paper (320, 400, 600, 3 μm, .3 μm grits) on a firm flat surface.

(h) Work from coarsest to finest grit and polish the end of the fiber and ferrule down to the surface of the polishing bushing. See Figure 7. Hold bushing firmly and alternate among circular, elliptical, and figure-8

patterns while polishing. Gently brush off face of polishing surface before starting any finer grit paper. Polish glass terminations under running water to flush away any silica chips that may form.

NOTE: *Do not use fine grit lapping films beyond their normal life span.*

(i) The polishing is completed when the surface of the polishing bushing immediately surrounding the ferrule is polished to a glossy finish. Do not polish excessively. See Figure 8.

(j) Discard the polishing bushing. Do not reuse.

3. CONNECTOR ASSEMBLY

(a) I/O Bushings — Mount bushing on panel. Insert source of detector device into slot and secure. Thread retaining cap onto bushing and hand tighten.

NOTE: *When inserting a source or detector device into a press-fit type bushing be sure the device is completely bottomed. The back end of an unterminated ferrule may be used as a tool to insert device.*

(b) Splice Bushings — Mount bulkhead bushings on panel. Thread retaining caps onto bushing and hand tighten.

Figure 4

Figure 5

Figure 6

Figure 7

Figure 8

Figure 9

Figure 7–82 Optimate Single Position Fiber-Optic Connector Installation (Sheet 2 of 2) (*Courtesy AMP Inc.*)

216

HOUSING	POS.	FUNCTION	MOUNTING	P/N
PLUG	2	SPLICE	FREE-HANGING	226952-1
	4	SPLICE	FREE-HANGING	226952-2
RECEPTACLE	2	SPLICE	PCB	226956-1
	4	SPLICE	PCB	226957-1
	2	TO-18 PACKAGE	PCB	227016-1
	2	TO-18/SPLICE	PCB	227017-1
	2	SPLICE	PCB, SNAP-IN*	226961-1
	2	SPLICE	FREE-HANGING	227018-1

* For use with .055" thick P.C. board only.

Figure 1

1. INTRODUCTION

The AMP Optimate Multiple Position Fiber Optic connectors join up to four optic cables in a self-centering splice, using color-coded ferrules which control the alignment parameters. Receptacles that accept TO-18 source or detector packages are also available.

2. SELECTION DATA

Select the proper plugs and receptacles from the Chart in Figure 1. Figure 2 matches the ferrule color coding with the corresponding cables. Part numbers of the other items needed to complete a connector are listed in Figure 3.

3. CABLE PREPARATION

(a) Carefully strip 1/2" of jacket from end of cable. Do not damage fiber. Strip glass bundles with hot wire strippers. Remove all strength members.

(b) Check to be sure fiber and ferrule are free of all contaminants and lubricants. Clean glass ultrasonically in alcohol, be sure no loose pieces of stranding remain in bundle.

★ Trademark of AMP Incorporated

FERRULE ASSEMBLY KIT			CABLE	
KIT P/N	FERRULE P/N	COLOR	JACKET DIAMETER	FIBER DIAMETER
227072-2	226953-2	RED	.090 IN./2.29 MM	.045 IN./1.14 MM
227072-3	226953-3	ORANGE	.077 IN./1.96 MM	.077 IN./1.96 MM
227072-4	226953-4	YELLOW	.090 IN./2.29 MM	.019 IN./0.48 MM
227072-5	226953-5	GREEN	.098 IN./2.49 MM	.024 IN./0.61 MM
227072-6	226953-6	BLUE	.062 IN./1.57 MM	.016 IN./0.41 MM
227072-7	226953-7	VIOLET	.062 IN./1.57 MM	.019 IN./0.48 MM
227072-8	226953-8	WHITE	.102 IN./2.59 MM	.016 IN./0.41 MM
1-227072-0	1-226953-0	BLACK	.078 IN./1.98 MM	.016 IN./0.41 MM

Figure 2

PART	P/N
CRIMP RING†	226954-1
COMPRESSION SPRING †	226955-1
POLISHING BUSHING†	530522-1
HAND TOOL	90364-1
2 POSITION STRAIN RELIEF**	226951-1
4 POSITION STRAIN RELIEF**	226951-2
2 POSITION BACKPLATE	226950-1
4 POSITION BACKPLATE	226950-2

** 2 halves required per assembly.
† Included in ferrule assembly kit P/N 227072-☐ along with ferrule P/N 226953-☐.
See Figure 2.

Figure 3

Figure 7–83 Optimate Multiple-Position Fiber-Optic Connector Installation Procedures (Sheet 1 of 2) (*Courtesy AMP Inc.*)

(c) Push crimp ring onto ferrule and slide compression spring over cable. See Figure 4.

(d) For glass bundles: Apply one drop of glue to fiber bundle. Gently fan out fibers and thoroughly mix glue into the bundle. Apply one drop of glue to the jacket. Insert cable into ferrule as far as it will go. See Figure 4.

NOTE: *Glues recommended for glass fibers are LOCTITE† 414, LOCTITE 495, or equivalent.*

For plastic bundles: Apply one drop of solventless epoxy glue (2 part, 100% solids) to the fiber and one drop to the jacket. Insert cable into ferrule as far as it will go. See Figure 4.

CAUTION: USE ONLY SOLVENTLESS, 2 PART, 100% SOLIDS, EPOXY GLUE. Any other type will damage fiber cladding.

Figure 4

(e) Position crimp ring against shoulder of ferrule. See Figure 5. Crimp with hand tool P/N 90364-1. Refer to IS 7885 for tool operation and maintenance.

(f) Push polishing bushing onto front of ferrule until ferrule is firmly bottomed. Tip of ferrule will protrude .020" (approx.) from bushing. See Figure 6.

(g) Carefully trim fiber so that it protrudes 1/16" (approx.) from tip of ferrule, as shown in Figure 6. Do not break fiber.

(h) Place 5 grades of silicon carbide finishing paper (320, 400, 600, 3 μm, .3 μm grits) on a firm flat surface.

(i) Work from coarsest to finest grit and polish the end of the fiber and ferrule down to the surface of the polishing bushing. See Figure 6. Hold bushing firmly and alternate among circular, elliptical, and figure-8 patterns while polishing. Gently brush off face of polishing surface before starting any finer grit paper. Polish glass terminations under running water to flush away any silica chips that may form.

NOTE: *Do not use fine grit lapping films beyond their normal life span.*

Figure 5

† Trademark of Loctite Corp.

(j) The polishing is completed when the surface of the polishing bushing immediately surrounding the ferrule is polished to a glossy finish. Do not polish excessively. See Figure 6.

(k) Discard the polishing bushing. Do not reuse.

Figure 6

4. CONNECTOR ASSEMBLY

(a) Mount receptacle on PCB. The "snap-in" type just pushes into position. The standard mount units are secured by self-threading nuts or by heat-peening the legs in place. See Figure 7.

(b) Insert completed ferrule assemblies into receptacle and plug. See Figure 8.

(c) Attach backplate or strain relief to rear of housings to secure ferrules. (Strain relief must be assembled by snapping the 2 pieces together around cables.)

(d) Join plug to receptacles by pushing together until latches at sides engage.

Figure 7

Figure 8

Figure 7–83 Optimate Multiple-Position Fiber-Optic Connector Installation (Sheet 2 of 2) (*Courtesy AMP Inc.*)

FIBER TESTING

As with any industry product, testing of optical fiber is accomplished to ensure that the fiber qualifies under the requirements of specification. Usually tests fall within the major headings of mechanical, optical, electrical, chemical, and physical. This does not preclude the possibility of a company performing a special test peculiar to its own fiber. Tests provide empirical or historical data that help to improve the product or production cost. Tests are performed under controlled temperature, humidity, and atmospheric conditions so as to obtain the same results. Other conditions such as cleanliness, lighting, accuracy of equipment, and personnel proficiency are considerations to be met.

There are literally dozens of tests that are performed on optical fibers. Since there is not space to describe all of these, a sampling of several mechanical and optical tests will be discussed here.

Fiber Tensile Strength   Tensile strength of a fiber is an extremely important parameter. It allows the designer of optical fiber systems more flexibility in the choice of components in the system. Further, it provides cable manufacturers with information that may afford the designer a margin of selection for the strength members within the cable. These are opportunities that behoove fiber engineers to develop a method of testing fiber for tensile strength.

Tensile strength of a fiber is governed by stress concentrations along the fiber. It appears from experience that surface flaws are principally responsible for high stress concentrations. Failure usually occurs at the deepest surface flaw. Empirical data also have shown that moisture imposed on the fiber surface (and therefore the flaws) tend to enlarge the flaws and cause stress corrosion. Stress corrosion has a fatigue limit that may cause premature failure.

Accurate measurement of tensile strength is complicated by the fact that fibers are usually long. Therefore testing involves checking of tensile strength for selected or random points on the fiber. By this method, large but rare flaws may be detected and then studied to determine their causes. Once the rare flaws are known, a sophisticated system of testing may be employed.

Proof Test   Each fiber manufacturer has a method of checking rated fiber tensile strength. Figure 7–84 illustrates a test set up whose principle of operation parallels tensile strength testing in the industry. An apparatus is set up with four equal diameter pulleys (A, B, C, D). The pulleys are mounted on axles on a strong mounting structure. Belt-driven motors provide rotation. Pulley A is motor-driven at a constant speed. A magnetic clutch on pulley B provides for tensile loading and rapid disengagement. The other two pulleys, C and D, are idlers.

A fiber specimen is strung through the constant-drive pulley A and the idler pulley C, and then the clutch pulley B and idler pulley D. Tension level is set by adjusting torque tension pulley B to a predetermined proof load. A friction wheel within the clutch of pulley B allows it to load-release quickly.

The entire length of the fiber is fed progressively through the setup while subjected to the constant tensile load for a period of time determined by the speed of drive pulley A and the distance between the centerlines of pulleys A and B. On the figure this area is called the region of tension. After passing over pulley B and D, the fiber is wound on a takeup reel with low tension.

Testing parameters are proof load, time under load, length of fiber under load, and number of breaks, if any.

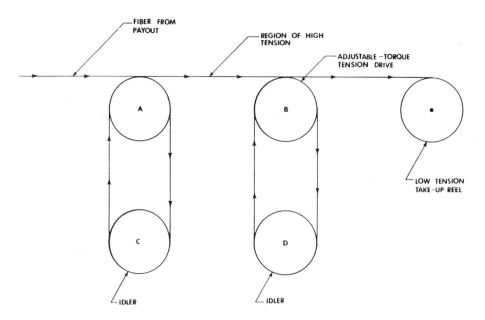

Figure 7–84 Proof Test (*Courtesy ITT Cannon Electric*)

Dynamic Short-Length Strength Test As you may recall, the proof test called for the testing of rated tension that should not break the fiber. The dynamic short-length strength test has tension applied until the fiber fractures. In figure 7–85, a length of fiber (say 2 meters) is placed between spools. The top spool is a holding spool that is fixed to a tension gauge. The lower spool is a constant-speed drive consisting of a threaded shaft, drive motor, a power supply, and a pulley attached to the threaded shaft. The power supply drives the motor, which turns the pulley. The pulley traverses the threaded shaft at a constant rate. The entire apparatus is attached to a rail for strength.

A predetermined tensile load is applied through the lower spool-drive mechanism at a specified speed until the fiber fractures. The fracture load, speed, and diameter of the fiber ends are monitored and recorded. Sample lengths from production runs are tested to accumulate an average for statistical and historical data. Data accumulation includes average tensile strength, load rate, fiber identification, and the number of samples.

Figure 7–85 Dynamic Short Length Strength Test (*Courtesy ITT Cannon Electric*)

Static Short-Length Strength Test The difference between static and dynamic tests is obvious. The fiber moves in a dynamic test and is static in a static test. As in the dynamic test, loads are applied to the fiber until it fractures. In figure 7–86, a specimen of fiber is placed between two pulleys under tension to prevent sagging. The pulleys are of equal diameter. A bracket on the right pulley in the figure is mounted to holding posts and/or tension gauges. One or the other pulleys have a ratchet jack for application of tension. The fiber ends are stripped of their jackets. A light source is applied to one end of the fiber and a convenient observation source on its other end. Two gauge markers and a measuring tape are placed on the cable (fiber) between the two pulleys.

Tension is applied to the fiber by operating the ratchet jack. Tension is applied at predetermined increments and monitored on the tension meter. The fiber length between gauge markers is measured at each increment until the fiber fractures. Significant parameters are tensile load, fiber elongation, and load at fracture. Fiber load is 50 percent of the tension gauge indication.

Figure 7–86 Static Short Length Strength Test (*Courtesy ITT Cannon Electric*)

Attenuation Test   Fiber attenuation is the ratio of power loss in decibels (dB) to the length of the fiber. The test setup for fiber attenuation is illustrated in the figure 7–87. The light source is a standard microscopic tungsten-filament bulb mounted within an illuminator housing. By adjustment of focusing optics a collimated beam is launched through a filter. Collimating is making beams parallel or straight. A slide tray of filters covering a variety of wavelengths from say 40 to 100 microns in length and in 0.01, 0.02 or 0.04-micron steps drops filters into a slide projector for filtering action. The filtered-collimated beam is directed into an optical chopper, where it is mechanically chopped to a reference signal of, say, 84 hertz. This chopped reference signal is coupled into a lock-in amplifier. The purpose of the reference signal is to screen out unwanted background light and to improve test sensitivity.

The chopped, filtered, and collimated beam is focused into a large-diameter plastic-clad silica fiber that is used as a point source for the fiber under test. The source fiber is attached to optical apparatus mounted on a vibration-free granite

Figure 7–87 Attenuation Test (*Courtesy ITT Electro-Optical Products Division*)

slab. The signal is again collimated through a set of television cameras with variable f-stops. The variable f-stops of the cameras allow the fiber under test to be illuminated by specific numerical apertures (NAs). The fiber end is positioned (aligned) with the light beam by a 5-axis micropositioner. This establishes maximum light launching into the fiber.

The test fiber output is adjusted by a 3-axis micropositioner to ensure maximum light beam to fall on light-shielded, high-quality photodiode (PIN) detector. Leads from the PIN diode are connected to the lock-in amplifier. The test fiber is prepared by stripping its jacket, scribing and breaking the fiber perpendicular to its core, and cleaning. Long length attenuation testing now takes place.

The fiber input end is placed into the 5-axis micropositioner. Bandpass filters are selected for the fiber type and placed into the source optics. Focusing optics adjust for maximum NA and maximum output. The signal output is then compared with a reference signal and the difference voltage signal read from the lock-in amplifier display. Note that the signal readout is a voltage signal that represents fiber attenuation.

Data is collected at each specific filter and NA setting by reading voltage value levels on the lock-in amplifier display. A baseline may be established by reading the lock-in amplifier display with the light beam blocked. Short-length testing of fiber may be compared by cutting a length of fiber off the injection end and preparing it, then installing it back in the test setup. No adjustment of the test setup should be made. Readings from the lock-in amplifier are again taken. Long-length and short-length fiber test results are then used to establish the attenuation coefficient.

Dispersion Test   Dispersion, as you may recall, is the spreading of ray arrival times at a receiver. Bandwidth reduction increases with length and is usually expressed in megahertz per kilometer. For pulse signals the effect results in pulse spreading which is generally expressed in nanoseconds per kilometer. Testing of pulse dispersion is readily explained. In the figure 7–88, a timing control unit provides a trigger pulse to the external trigger (horizontal) of a sampling oscilloscope. The timing control unit consists of an integrated-circuit clock triggered by a low-frequency oscillator. A second pulse from the timing control unit triggers a laser driver. A step time delay in, say, 100 nanosecond steps is transmitted to the oscilloscope.

A laser diode is driven by the laser driver. A laser trigger pulse from the timing control pulse initiates the operation. The laser produces a narrow light bandwidth into the optical setup. The optical system (in the center of the illustration) consists of four microscope-objective lenses arranged on a square plate. A beamsplitter is aligned at a 45° angle with respect to the optical axis on the square plate as shown in the figure. Light directed from the laser diode is collimated at objective lens A. The beamsplitter reflects part of the beam into objective lens D where it is focused onto the PIN detector (photodiode) as a reference level. The remainder of the light is transmitted through the beamsplitter and focused through objective lens B and into the fiber under test. The opposite end of the fiber under test is attached at objective lens C where the output light is recollimated and directed through the beamsplitter and focused by objective lens D into the PIN detector.

An APD (avalanche photodiode) detector is mounted external to the test setup. Before the test, the fiber end is installed on the APD detector. The APD detector output is displayed on sampling oscilloscope for initial alignment, by manipulation of the input end micropositioner. Long-length measurements are begun. The fiber is removed from the APD detector and placed on the objective lens C. The timing control unit is originally set to 0.0 delay then in successive increments of 0.1 microseconds until the pulse is observed on the sampling oscilloscope. The fiber input micropositioner is fine-adjusted to get a maximum output picture on the oscilloscope.

Laser driver current is adjusted just above threshold and then at threshold. The traces of the above threshold and at threshold laser operation are recorded on the x-y chart recorder.

Figure 7–88 Dispersion Test (*Courtesy ITT Electro-Optical Products Division*)

Short-term measurements are then taken. The fiber is shortened by several meters and reprepared. Alignment is made and traces obtained in the same manner as long-length measurements. Peak signal current is determined by measuring the maximum difference between laser threshold and the pulse waveform on the x-y chart recording. This current difference is then multiplied by 0.5 to determine the −3-decibel pulse height. You may recall from past electronic experience that the −3-decibel point is that point where power loss has fallen to half or actually 0.501.

The half-power pulse height is then marked on the leading and trailing edges of the pulse waveform, using the baseline voltage as a reference. Pulse-width is measured between the leading- and trailing-edge cross marks.

Numerical Aperture (NA) Measurements Using the Scanning Method A shelf is installed on a rotating stage. The fiber output end is taken from the attenuation test illustrated in figure 7–89 and attached to the shelf. The shelf is marked so as to be able to bring the fiber endface parallel to the detector at 0° rotation angle of the rotating stage. A large-area detector is positioned about 40 centimeters from the fiber output face. This test setup is identical to the attenuation test setup

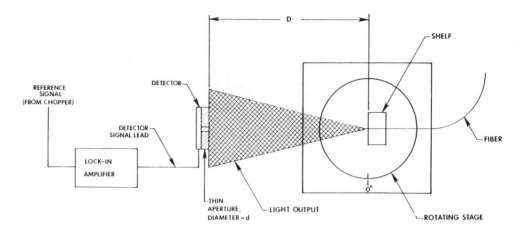

Figure 7–89 Numerical Aperture (NA) Test Setup (Scanning Method) (*Courtesy ITT Electro-Optical Products Division*)

just discussed and illustrated in figure 7–87, with minor variations. The rotating stage is placed between the output end of the fiber under test and the detector.

A reference signal is placed into the lock-in amplifier and the detector leads are attached to the lock-in amplifier. The detector is set so that it collects maximum signal at 0° rotation of the stage. To measure the fiber output pattern the rotating stage is adjusted to an angle which produces zero signal (less than 10 percent) at maximum full-degree angle (example 16° rotation). A horizontal line is drawn on graph paper. This line represents the distance (40 centimeters) between the fiber output face and the detector. It also represents the maximum full-degree angle which just produced a zero signal (less than 10 percent).

The rotating stage is then turned in the opposite direction in 1° increments and detector voltage readings are taken at each stop until the beam limit is reached on the opposite side of the beam pattern. That is the maximum full-degree angle which produces a zero signal (less than 10 percent). A line is drawn on the graph paper from the fiber endface (the right side of the horizontal line) at an angle equal to the difference between the two maximum zero-signal angles. These are determined by the rotation of the shelf. The fiber NA can then be calculated simply by taking the sine of half the angle between the two zero signals (less than 10 percent points).

A second method for graphing is to draw a horizontal line that represents the 40-centimeter distance between the fiber face and the detector. Draw a second line vertically in each direction on the detector end of the horizontal line. Indices on the vertical line should represent lock-in amplifier voltage output as related to angle of shelf rotation. Maximum full-degree angle voltages are plotted on the graph vertical as read on the lock-in amplifier. These zero-signal points are connected to the fiber end of the horizontal line. One-half the angle made by the two extremes is used to calculate the NA. That is, NA = sine ½ the angle between the two extreme zero-signal points.

Numerical Aperture (NA) Measurements Using the Trigonometric Method

The test setup is mounted on a vibration-isolated granite slab (see figure 7–90). The fiber under test is prepared in the normal manner. The input end of the test fiber is placed on a 5-axis micropositioner. A wide-angle light source launches a beam of light through optics into the test fiber. The opposite end (output) is placed in front of an image screen. The screen is positioned a known distance (D) from the fiber endface so that the output light spot falls on the screen. Alignment of the fiber and the input optics is made so that the output spot is maximum size and intensity. A vernier caliper is used to measure the spot size.

Calculation of the numerical aperture (NA) is made with the results using the following formula:

$$NA = \frac{A}{\sqrt{A^2 + 4D^2}}$$

where   A = the spot diameter
D = the distance from fiber to screen

Figure 7–90   Numerical Aperture (NA) Test Setup (Trigonometric Method) (*Courtesy ITT Electro-Optical Products Division*)

Some General Test Procedures Used in Fiber-Optics Industry

Test procedures are varied and range from simple to complex. The test procedures listed here are just a small number of ones that are in actual use. Each manufacturer will also have its own method or technique to perform the test for best results and speed of accomplishment.

Fiber Size Measurement   This test measures the core and cladding of an optical fiber. The test is accomplished using a microscope, a prepared fiber, and a predetermined calibration index.

Fiber Bundle Measurement   The measurement checks bundle size by dipping the bundle into wetting liquid, then placing the bundle into sizing gauges beginning with the largest gauge.

Number of Fibers   Determination of the total number of fibers in a fiber bundle where the number of fibers is large. The fiber bundle end is prepared and photographed. The photo is enlarged to applicable size and the fibers are counted.

Number of Transmitting Fibers   Test is used to determine the number of fibers in a bundle that are transmitting light. Several photographs are taken using different intensity settings. Photos are enlarged as necessary and the fibers transmitting are counted.

Cyclic Flexing   A procedure used to determine the ability of a fiber-optic cable to withstand cyclic flexing. The fiber is placed in a cyclic arm driven by an electric motor. The cyclic arm flexes the fiber a specified count.

Low Temperature Flexibility   A test which determines the ability of a fiber-optic cable to withstand bending around an obstacle at low temperature by measuring fiber breakage or transmitted power. The fiber is bent around mandrels, then straightened while under cold temperatures.

Impact Testing   A procedure used to determine the ability of a fiber-optic cable to withstand impact loads. A drop hammer of specified weight free falls on a length of fiber during this test to determine damage.

Compressive Strength   A test to check the ability of a fiber-optic cable to withstand slow compression or crushing. A fixture that applies compression to the fiber is used and an inspection is made to determine damage.

Cable Twist and Bend   Test is used to determine the ability of a fiber-optic cable to withstand twisting checks are made for broken fibers and/or attenuation. A pair of gripping blocks hold a length of fiber. One of the blocks is fitted with a rotating clamp to twist the fiber. Fiber is then inspected for damage.

Cable Tension Load   This test is used to determine percent increase in length under tension, broken fibers, changes in radiant power, and the ability of a strength member to withstand tensile loads. Load is placed on the fiber in a form such as gripping blocks. The fiber is placed under tension. Measuring gauges check tension and length.

Power Transmission versus Temperature   The effects of temperature upon the transmitted power of an optical cable are defined by this test. This is a relation between two temperatures. A light source and detector are placed on either side of a test chamber. Transmission is made while temperature is varied.

Power Transmission versus Temperature Cycling   A test to measure the effects of temperature upon the transmitted power of an optical cable while cycling temperatures. Fiber is placed in high then low test chambers. The light source and detector are on the outside of the test chamber. The fiber is exposed in cycles to high and low temperatures.

Power Transmission versus Humidity   This test determines the effect of humidity upon transmitted power. This is a ratio of radiant power of initial humidity preconditioning and at a final condition. A fiber specimen is placed in a humidity-controlled test chamber with the light source and detector externally mounted. Control is made to ensure condensation does not fall on the fiber.

Tensile Loading versus Humidity   This test defines the effect of humidity on the tensile loading of a fiber-optic cable. Test is performed within a test chamber. Loads are placed on the fiber under controlled humidity.

Freezing Water Immersion—Ice Crush   A test to determine the effect of crush force caused by freezing water upon the transmitted power of an optic cable immersed in freezing water. This is a ratio of radiant power in cold but unfrozen water and in frozen water. Fiber is placed in a test chamber that is able to lower temperatures quickly at a predetermined rate.

Dimensional Stability   This test determines the permanent dimensional changes which occur to a fiber jacket or covering when the cable is exposed to elevated temperatures. The test checks expansion and shrinkage usually in a percent of change. The test fiber is dimensionally checked before and after exposure to heat in a test chamber.

Flammability   The test measures the flammability of a subject fiber by application of a flame such as a bunsen burner, then checking for fiber damage.

Far-End Crosstalk   The test measures the crosstalk of two neighboring transmitting fibers. Two fiber specimens are placed parallel to each other, radiation is applied to one fiber and detectors are used to monitor radiation emanating from the illuminated and neighboring fibers.

Refractive Index Profile   This test determines the refractive index profile of a graded-index fiber by the interferometric process. The fiber end is prepared and placed under a microscope. The microscope is tilted so as to view a tilted wavefront on the fiber face. A photo is taken for empirical data.

Military Specifications and Standards   Test procedures for fiber optics fall into many categories, the main ones are general physical, electrical, and chemical methods. Most of these standards and specifications are referenced to federal and military documents.

SPECIFICATIONS

Federal

L-P-390	Plastic molding material, polyethylene, low and medium density
QQ-W-343	Wire, electrical and nonelectrical, copper, (uninsulated)
QQ-W-423	Wire, steel, corrosion-resisting

Military

MIL-C-17	Cables, radio frequency; coaxial, dual coaxial, twin conductor and twin lead
MIL-I-631	Insulation, electrical, synthetic-resin composition, nonrigid
MIL-W-5086	Wire, electric, hookup and interconnecting, polyvinylchloride-insulated, copper or copper alloy conductor
MIL-C-12000	Cable, cord, and wire, electric, packaging of
MIL-C-13777F	Cable special purpose, electrical
MIL-M-20693	Molded plastic, polyamide (nylon) rigid

STANDARDS

Federal

FED-STD-191	Textile test methods
FED-STD-228	Cable and wire, insulated; methods of testing
FED-STD-601	Rubber, sampling and testing

Military

MIL-STD-104	Limits for electrical insulation color
MIL-STD-109	Quality assurance terms and definitions
MIL-STD-202	Test methods for electrical and electronic components
MIL-STD-810	Environmental test methods

8

FIBER-OPTIC SYSTEMS AND APPLICATIONS

GENERAL

This chapter is dedicated to putting the optical fiber to use. In the first part of the chapter we will discuss some of the things that are involved in organizing the system. In the rest of the chapter, functional and unique system applications for use of fiber optics are presented.

A BASIC FIBER-OPTIC SYSTEM

A simple fiber-optic system is called a *transmission link*. It consists of a transmitter with a light source, a length of fiber, and a receiver with a light detector. The basic operation of a system is to connect a digital or analog signal to a transmitter. Within the transmitter, the input signals are converted from electrical to optical energy by modulating an optical light source, normally achieved by varying the drive current. The modulated light is launched into a length of fiber where it reflects from wall to wall through the fiber core. At the opposite end of the fiber a detector accepts the light and converts it back to an electrical signal. The electrical signal is converted back to its original form in the receiver. A brief discussion of each major component is provided.

Electrical Signal Transmitter (Driver) The purpose of the transmitter (driver) is to change the electrical signal into the required current to drive a low-impedance light source. The electrical inputs are either digital or analog. The choice of the converter should depend on the current requirement of the light

231

source. If the signal is digital, the transmitter (driver) should consist of a high-speed pulser to turn the light source on and off. If the signal is analog, the transmitter (driver) should be able to supply current to the light source to transmit the positive and negative alternations of the signal.

In figure 8–1, a typical analog driver is illustrated. R_2 and R_3 provide a voltage divider for the input signal. The potentiometer P_1 and resistor R_1 serve to set the operating point so that the positive and negative swings of the analog input signal produce only a positive output. The output current never changes polarity, only amplitude. R_{FB} is the feedback resistance. The light-emitting diode (LED) transmits light as current varies. Resistor R_L provides current limiting and load.

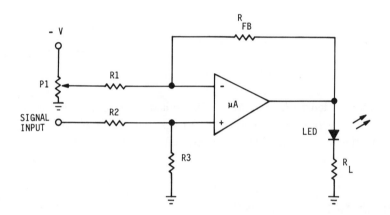

Figure 8–1   A Typical Analog Driver (*Mike Anaya*)

In figure 8–2A, a very basic LED driver consists of an inverter with an LED tied to its output. The resistor R_L serves as a current limiter. Signal pulses at the input direct pulse current to flow through the LED causing it to radiate.

In figure 8–2B, a variable (analog) signal input is directed through a pulse modulator. The modulator can be one of three types: PRM—pulse-rate modulation, PPM—pulse-position modulation, or PWM—pulse-width modulation. Modulated output directs pulse current to flow through the LED causing it to radiate.

The samples in figures 8–1, and 8–2, are extremely simplified. More details as to transmitter (driver) parameters and operation are discussed later on in this chapter.

Light Source   The purpose of the light source is to launch a light signal into the optical fiber at an angle that provides maximum signal transfer. There are

(A) INVERTER DRIVER.

(B) PULSE MODULATED DRIVER.

Figure 8–2 Basic LED Drivers (*Mike Anaya*)

two basic light sources used in fiber-optic electronics. These are the light-emitting diode (LED) and the injection laser diode (ILD). Both of these units provide small size, brightness, low drive voltage, and are able to emit signals at desired wavelengths. Each has characteristics that make them desirable or undesirable for a particular application. The LED has a longer lifespan, greater stability, wider temperature range, and much lower cost. The ILD is capable of producing as much as 10 decibels more power output than the LED. It can launch the light signal at a much narrower numerical aperture (NA), and therefore can couple more power through the optic fiber than the LED. The disadvantage of using an ILD is that its current range is extremely restricted. Since some system operations vary greatly, the ILD must have compensation devices added to the electronics. This may make the cost prohibitive. More details as to light-source parameters may be found in chapter 6.

Light Detector   The light detector accepts a light signal from the optical fiber and converts it into an electrical current. There are three types of detectors in use: the phototransistor, the PIN diode, and the APD.

The phototransistor is inexpensive but has slow rise times and a limited bandwidth capability. Therefore it is seldom used.

The PIN diode (contains positive, intrinsic, and negative solid-state layers in its construction) exhibits fast rise time and acceptable bandwidth parameters. It is reasonably priced.

The APD diode (avalanche photodiode) exhibits fast rise time and acceptable bandwidth parameters. The APD is more expensive than the PIN diode since it provides greater receiver sensitivity. The APD also requires an auxiliary power supply. More details on the PIN and APD diodes are furnished in chapter 6.

Receiver   The function of the receiver is to accept low-level power from the detector and convert it into a high-voltage output. There are at least two methods of accomplishing this. In figure 8–3A, detector current produces a voltage drop across a load resistance. The voltage drop is directed into an amplifier. An output voltage representative of the transmitted signal is the result. In figure 8–3B, the operational amplifier output voltage is the effect of the amplifier driving detector current through the feedback resistance. Again the output voltage is representative of the transmitted signal.

Other electronics may be added to the circuitry to maintain correct response. A gain control may be used on the front end to vary the impedance of the receiver. The operational amplifier is used as a current to voltage converter.

Since signal inputs are generally weak, shielding and power-supply decoupling are a requirement to achieve sensitivity. Sensitivity is set by input noise. Noise, of course, leads to errors in digital systems and restricts signal-to-noise ratio (SNR) in analog systems. Details on optical receivers are discussed later on in this chapter.

SYSTEM COMPATIBILITY AND SPECIFICATIONS

It stands to reason that if you acquire all the parts for a fiber-optic system from one company, the parts you buy will be compatible. Indeed, it may be intelligent thinking to do just that. Most manufacturers have developed complete systems and have off-the-shelf components available for the asking. Let's consider some

(A) DETECTOR CURRENT CAUSING VOLTAGE DROP ACROSS A LOAD RESISTOR.

(B) AMPLIFIER DRIVES DETECTOR CURRENT THRU FEEDBACK RESISTOR.

Figure 8–3 Receiver Action (*Mike Anaya*)

of the things that are ultimately important when parts are chosen and a system is being created.

System Considerations
1. Digital
   a. Required bit error rate (BER) in bits per second. Upper bit error rate is usually in megabits per second (Mb/s). Lower bit error rate is usually in bits per second (b/s)
   b. Temperature operating range in degrees centigrade (°C)
2. Analog
   a. Bandwidth in hertz (Hz), or megahertz (MHz)
   b. Distortion in decibels (dB) or %
   c. Temperature operating range in degrees centigrade (°C)
3. Audio
   a. Bandwidth in hertz (Hz) or megahertz (MHz)
   b. Distortion in decibels (dB)
   c. Crosstalk in decibels (dB) (for multiple channels)
   d. Temperature operating range in degrees centigrade (°C)
4. Video
   a. Bandwidth in hertz (Hz) or megahertz (MHz)
   b. Distortion in decibels (dB)
   c. Crosstalk in decibels (dB) (for multiple channels)
   d. Temperature operating range in degrees centigrade (°C)

Transmitter Specifications
1. Input impedance in ohms (Ω)
2. Maximum input signal in volts dc (V_{DC}), volts effective (V_{RMS} V_{EFF}) volts peak-to-peak (V_{P-P})
3. Optical wavelength in micrometers (μm) or nanometers (nm)
4. Optical output power in microwatts (μW)
5. Optical output rise time in nanoseconds (ns)
6. Required power supply in volts dc, usually 5 ± 0.25 V_{DC} or $\pm 15 \pm 1$ V_{DC}

Light-Source Specifications
1. Continuous forward current in milliamps (mA)
2. Pulsed forward current in milliamps (mA)
3. Peak emission wavelength in nanometers (nm)
4. Spectral width in nanometers (nm)
5. Peak forward voltage in volts dc (V_{DC})
6. Reverse voltage in volts dc (V_{DC})
7. Temperature range in degrees centigrade ($^\circ$C)
8. Total optical power output in microwatts (μW)
9. Rise/fall times in nanoseconds (ns)

Fiber Specifications
1. Mode—single or multimode
2. Index—step or graded
3. Attenuation in decibels per kilometer (dB/km)
4. Numerical aperture (a sine value)
5. Intermodal dispersion in nanoseconds per kilometer (ns/km)
6. Core index of refraction (a ratio)
7. Cladding index of refraction (a ratio)
8. Core diameter in micrometers (μm)
9. Cladding diameter in micrometers (μm)
10. Tensile strength in pounds per square inch (psi)
11. Bend radius in centimeters (cm)

Cable Specifications
1. Numbers of fibers (a unit)
2. Core diameter in micrometers (μm)
3. Cladding diameter in micrometers (μm)
4. Cable diameter in millimeters (mm)

5. Weight in kilograms per kilometer (kg/km)
6. Minimum bend radius in centimeters (cm)

Detector Specifications
1. Continuous forward current in milliamps (mA)
2. Pulsed forward current in milliamps (mA)
3. Peak reverse voltage in volts dc (V_{DC})
4. Temperature range in degrees centigrade (°C)
5. Optical power output in microwatts (μW)
6. Threshold current in milliamps (mA)
7. Rise/fall times in nanoseconds (ns)
8. Radiation pattern in angular degrees

Receiver Specifications
1. Output impedance in ohms (Ω)
2. Output signal level in volts dc (V_{DC}), volts effective (V_{RMS} or V_{EFF}), volts peak to peak (V_{P-P})
3. Optical sensitivity in microwatts (mW), nanowatts (nW), or decibels (dB), or megabits per second (Mb/s)
4. Optical dynamic range in decibels (dB)
5. Analog output overload (in %)
6. Analog output rise time in nanoseconds (ns)
7. Digital output rise time in nanoseconds (ns)
8. Required power supply in volts dc, usually $5 \pm 0.25\ V_{DC}$ or $\pm 15 \pm 1\ V_{DC}$

DESIGN CONSIDERATIONS

Before any thoughts of developing an optical fiber system can be completed, certain factors must be realized. First of all, the signal information must be known. Is the signal analog or digital? What is the information bandwidth? What power is required? Secondly, length of the transmission line has to be determined. How far is it between transmitter and receiver? Are there any physical obstacles that must be skirted or must the cable go through anything? Finally, what are the tolerable signal parameters? What is the acceptable signal-to-noise ratio (SNR) if the system is analog? What is the acceptable bit error rate (BER) and rise/fall time if a digital system? Once the basic parameters of the system are set, the system development can take place.

Design Procedures   Procedures for design of an optical fiber system are as follows:

1. Determine the signal bandwidth.
2. Determine the signal-to-noise ratio (SNR) if the signal is analog. This is the ratio of output signal voltage to noise voltage. The ratio is expressed as 10:1, 8:1, 20:1, etc. The largest signal-to-noise ratio is desirable. The SNR is expressed in decibels (dB). SNR curves are provided on detector data sheets.
3. Determine the tolerable bit error rate (BER) if the signal is digital. BER is the ratio of incorrect bits to total bits of data. A typical good BER is 10^{-8}. BER curves are supplied by the manufacturers of detectors.
4. Determine link distance, that is, the distance between the transmitter and the receiver.
5. Select a fiber based on attenuation.
6. Calculate fiber bandwidth for the system. This is accomplished by dividing the bandwidth factor in megahertz per kilometer by the link distance. The bandwidth factor is provided on fiber manufacturer's data sheets.
7. Determine power margin. This is the difference between light source output power and receiver sensitivity.
8. Determine total fiber loss by multiplying fiber loss in decibels per kilometer by the length of the link in kilometers.
9. Identify the number of connectors. Multiply the connector loss (provided by the manufacturer) by the number of connectors.
10. Identify the number of splices. Multiply the splice loss (provided by the manufacturer) by the number of optics.
11. Allow 1 decibel for detector coupling loss.
12. Allow 3 decibels for temperature degradation.
13. Allow 3 decibels for time degradation.
14. Sum the fiber loss, connector loss, splice loss, detector coupling loss, temperature degradation loss, and time degradation loss, (add values of steps 8 through 13) to find total system attenuation.
15. Subtract total system attenuation from power margin. If the difference is negative, then the light-source power receiver sensitivity must be changed to create a larger power margin. Secondly a fiber with a lower fiber loss may be chosen. Use of fewer connectors and splices may also be an alternative, if it is possible to do so without degrading the system.

16. Determine rise time. To find the total rise time add the rise time of all critical components such as the light source, intermodal dispersion, intramodal dispersion, and detector. Square the rise times. Then take the sum of the total squares. Square this sum and multiply it by some parameter factor such as 110 percent or 1.1 as in the following example.

$$\text{System rise time} = 1.1 \ \sqrt{T1^2 + T2^2 + T3^2 + \ldots}$$

Fiber-Optic System Attenuation The total attenuation of an optical fiber system is the difference between the power leaving the light source and the power entering the detector. In figure 8–4, power entering the fiber is designated as P_S or source power. L_{C_1} is power loss at the source to fiber coupling, usually 1 decibel per coupling. The power is of that signal launched into the fiber from the light source at the fiber coupling. L_{F_1} represents the loss in the fiber between the source and the splice. Fiber losses are listed in specifications and have power losses at about 10 decibels per kilometer. L_{SP} represents the power loss at the splice.

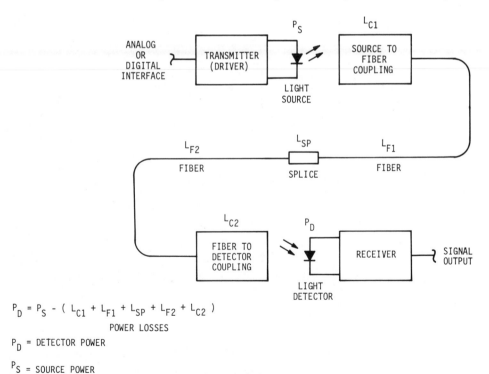

$$P_D = P_S - (L_{C1} + L_{F1} + L_{SP} + L_{F2} + L_{C2})$$
$$\text{POWER LOSSES}$$

P_D = DETECTOR POWER

P_S = SOURCE POWER

Figure 8–4 Fiber-Optic System Attenuation (Power Loss) (*Mike Anaya*)

A representative power loss of a splice is 0.3 to 0.5 decibels. L_{F_2} represents power loss in the second length of fiber. L_{C_2} is the power loss at the fiber-to-detector coupling. Finally, P_D is the power transmitted into the detector. Other power losses due to temperature and time degradation are generally around 3-decibels loss each. Power at the detector is then generalized as

$$P_D = P_S - (L_{C_1} + L_{F_1} + L_{SP} + L_{F_2} + L_{C_2})$$

Note: All power and losses must be expressed in decibels (dB).

   Transfer Functions   Transfer functions are the power formulas that describe the power at specific points along the transmission line. In figure 8–5, P_S is representative of the power output from the light source. Power at different points can be represented by a power formula called a *transfer function*. Some typical transfer functions applicable to the fiber-optic system in figure 8–5 are listed in table 8–1.

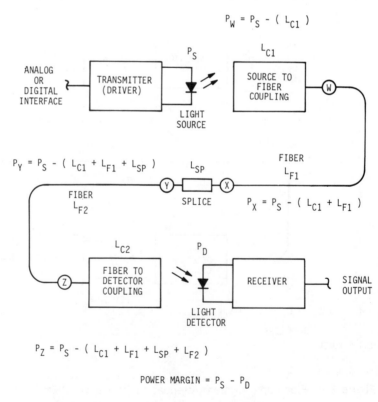

Figure 8–5   Transfer Functions (*Mike Anaya*)

TABLE 8–1 Table of Transfer Functions (See figure 8–5.)

Parameter	Transfer Function
Power out of light source	P_s
Power into fiber at point W	$P_W = P_s - L_{C1}$
Power into splice at point X	$P_X = P_s - (L_{C1} + L_{F1})$
Power into fiber at point Y	$P_Y = P_s - (L_{C1} + L_{F1} + L_{SP})$
Power into fiber at point Z	$P_Z = P_s - (L_{C1} + L_{F1} + L_{SP} + L_{F2})$
Power into detector	$P_D = P_s - (L_{C1} + L_{F1} + L_{SP} + L_{F2} + L_{C2})$

SYSTEM CABLING AND CONNECTORS

Although we previously discussed cables and connectors in chapter 7, it is appropriate to provide some discussion of them in the systems portion of this book. To replace electrical conductors, optical waveguide cables must attain mechanical and environmental capabilities comparable to the metal conductor cables. Acceptable cable structures must achieve thermal and elastic compatibility, and stress fibers to a controlled and uniform degree under both tensile and crush loading. They must prevent fiber-to-fiber contact, especially in a crossover configuration.

Typical Cable A fiber-optic cable that meets the requirements of realistic applications should consist of relatively few fibers, each mechanically protected. A typical fiber cable, illustrated in figure 8–6, consists of six channels (waveguides), each mechanically and optically isolated from the others. Note that the cable in the illustration is surrounded with a jacket and strength members. This particular cable has a maximum attenuation of 20 decibels per kilometer at 820-micron wavelength. The cable is 5 millimeters in diameter and has a tensile

Figure 8–6 A Typical Fiber Cable (*Courtesy Corning Glass Works*)

strength of 50 kilograms (110 pounds). Reference should be made to chapter 7 for other cable structures. Typical cables for specific systems will also be illustrated later in this chapter. Other cable designs incorporating various numbers of channels and graded-index waveguides are available.

Cable Connectors   In-line system connectors usually fall into two specific categories. The first has the fiber group butted collectively against the second group of fibers. The second butts individual fibers. In group connections losses in transmission are typically 3 to 5 decibels. The group connectors are easily made. The single-fiber connection has relatively low losses, in the neighborhood of 0.5 decibel, but are more expensive.

The desirable properties of waveguide connectors are limited by the bundle-to-bundle connectors. In addition to the severe insertion loss, grinding and polishing are required, and epoxy or other glue limits the connector performance, especially with regard to temperature, humidity, and chemical attack.

Individual fiber-to-fiber connectors will substantially eliminate the objections to the bundle-to-bundle connectors. However, well-matched fibers and precision connector hardware are required. In addition, care must be taken not to distort the fibers, which can cause significant scattering loss. This effect is encountered especially in connectors using adhesives (epoxies), due to material inhomogeneity or nonuniform curing. Finally, fiber ends must be smooth to avoid another cause of scattering losses. Fiber-end losses can be minimized through the use of index-matching fluids, which also reduce Fresnel reflection losses.

In addition to the need to minimize insertion loss in a structure mechanically and environmentally suited for its use and compatible with the cable, a connector should not require adhesive to hold the fiber, or grinding and polishing to finish the fiber end. Simplicity of connector hardware and of the termination process is important relative to future production costs and indispensable for practical field termination.

One connector concept locates the waveguide in the interstices of appropriately dimensioned rods of elastically suitable materials. (See figure 8–7.) It is immediately apparent that this approach has a self-aligning capability minimizing requirements for tight fiber tolerances. The rods are made of a compliant material for fiber retention without the use of adhesives. Due to crosstalk considerations, only every other interstitial space is used.

A second cable-connector concept is the overlap concept. The connector assembly technique allows simultaneous cutting of all waveguides based on controlled scoring of bent and stressed fibers. This technique has consistently viable perpendicular and fault-free fiber ends, eliminating the need for grinding and polishing.

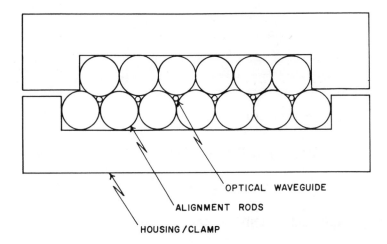

OPTICAL WAVEGUIDE

ALIGNMENT RODS

HOUSING / CLAMP

Figure 8–7 A Typical Cable Connector Concept (*Courtesy Corning Glass Works*)

TIGHT
CLAMP
REGION

SPLAYING
REGION

OPTICAL
WAVEGUIDES

CORGUIDE ™

CABLE
CLAMP
REGION

ACCUMULATION
REGION

Figure 8–8 Integration of Cable End into the Overlap Connector (*Courtesy Corning Glass Works*)

A viable connector concept must not impose the optical dimensional tolerances on the mechanical elements of the assembly. This is accomplished by separation of the mechanical and optical mating planes as shown in figures 8–8 and 8–9. In addition, the waveguides are made to function as springs, assuring contact of fiber ends. This is likely to prove particularly important under shock and vibration conditions over a broad operating temperature range.

Figure 8–9 The Overlap Connector (*Courtesy Corning Glass Works*)

TYPICAL SPECIALTY CABLE AND ITS APPLICATION

The type AT optical fiber cable is typical of the cables used throughout the communication industry. (See figure 8–10.) The AT cable was developed by General Cable Company and installed by General Telephone of California for use in what was reported to be the first use of optical fiber cable for regular telephone use. The cable was installed in a General Telephone route between the Artesia Central Office and the Long Beach Toll Center. Two regenerators (boosters) were placed in the line. The entire length of the system is 5.6 miles. One regenerator is 1.6 miles from the Long Beach Toll Center and the second regenerator 1.7 miles from the Artesia Central Office. The distance between the regenerators is 2.3 miles.

      AT Cable Construction A line drawing of the AT cable and its parts is illustrated in figure 8–11. A photograph of the AT cable is shown in figure 8–12. The optical fibers are laid parallel to each other and incorporated into a flat ribbon between plastic tapes that can be stripped back to expose the fibers when necessary. The ribbon is about 6 millimeters in width and contains up to 12 fibers. The ribbon construction reduces the effects of microbending and also provides additional strength as well as simplified and much safer handling of fibers during cable manufacture and in subsequent splicing operations.

      The ribbon containing the fibers is placed in a groove fashioned helically along the length of an extruded plastic core. The core is strengthened with a central copper wire that also serves as an electrical conductor. The cable sheath consists of a 1.9-centimeter diameter aluminum tube of 1.3-millimeter wall thickness constructed from a strip bent and longitudinally welded in a continuous process to form a loose fit over the cable core. Covering the aluminum tube is a poly-

Figure 8–10 A Typical Communication Fiber-Optic Cable Layout (*Courtesy GTE Corp.*)

ethylene jacket, followed by an 0.15-millimeter corrugated steel tape with longitudinal overlip, and, finally, a second and outer polyethylene jacket. The metallic surfaces are flooded with an asphaltic anticorrosion compound. The production processes are continuous and place no restriction on the length of the cable that can be manufactured. The overall diameter of the cable is approximately 2.5 centimeters.

Other Cable Types Figure 8–13 illustrates seven typical optical fibers used throughout the telephone and communications systems today. Each has special applications. From left to right in the photo, the following list describes some of their construction.

    a. Single-fiber patch cord, kevlar yarn, plastic jacket.
    b. Multiple fibers, water-block filling, plastic jacket.
    c. Multiple fibers, water-block filling, kevlar-reinforced plastic jacket.

Figure 8–11 Line Drawing of Type AT Cable (*Courtesy General Cable Company*)

Figure 8–12 The Type AT Cable (*Courtesy General Cable Company*)

    d. Multiple fibers, water-block filling, Bupic jacket.

    e. Multiple fibers, water-block filling, plastic jacket.

    f. Multiple fibers, water-block filling, inner jacket, FPA (fused polyethyl-ene-aluminum) outer jacket.

    g. Multiple fibers, water-block filling, kevlar-reinforced inner jacket, aluminum tube, outer jacket.

Attenuation of each fiber is a maximum of 8 decibels per kilometer. The attenuation average of all fibers in a reel length of cable is no more than 6.5 decibels per kilometer.

Figure 8–13 Typical Fiber Optic Communication Cables *(Courtesy General Cable Company)*

SYSTEM APPLICATION

Probably the most significant of the applications of fiber-optic systems are in the telephone industry. Substitutions for electrical conductors are being used throughout the country. Standardization within the industry is made possible by The North

American Digital Hierarchy, which is the standard in the United States. Essentially the standards are in channels and bit rates. These are listed below. Applications in telephone are intercity trunks, interoffice trunks, and local loops. The key advantages for telephone systems are high bandwidth, low attenuation, and small size.

System Line	4 KH$_z$ Voice Channels	Bit Rate
T1	24	1.544 M bit/sec
T2	96	6.312 M bit/sec
T3	672	44.736 M bit/sec
T4	4032	274.176 M bit/sec

Broadband networks such as television are using fiber waveguides in commercial cable television (CATV), wired city (community), and dedicated communication systems. The advantages are simpler transmission-line characteristics and noninterference (security).

Computers for general electronic data processing and computer-based process control and instrumentation systems represent large-scale usage of optical waveguides. At a given bit rate, the benefits of optical waveguides versus metallic conductors increase with distance. This, together with optical waveguides' immunity to electromagnetic interference, suggests the greatest motivation to apply waveguides in computer systems operating in a high-EMI environment. Computer applications for fiber optics are dispersed systems, intersystem wiring, computer-based industrial process control, and instrumentation. The advantage is immunity to interference.

Military applications include wiring of weapons systems such as aircraft, helicopter, ship, or submarine. Technically, many of them entail data transmission within computer-based systems. Headquarters, field, and base communications are of special importance to the Army. Surveillance systems operated from shore, ship, or submarine may make a substantial impact on the Navy's capabilities to detect, identify, and evaluate underwater threats. Secure communications like those for missiles are receiving attention. Having recognized the potential of optical waveguides very early, the military laboratories and their contractors have an impressive number of programs underway. These programs are yielding significant technical results and are expected to conclude with a line of military-qualified hardware, demonstrated system feasibility, and know-how to apply the technology of optical waveguides. Again, the advantages are immunity to electromagnetic interference (EMI) and electromagnetic pulses (EMP), security, and the size/weight factors.

LARGE FIBER OPTIC INSTALLATION

One of the major installations using fiber optics today is the Lake Placid Winter Olympics Lightwave System. The system was a joint effort of New York Telephone, AT & T, Western Electric, and Bell Labs. (See figure 8–14.) The new installation was obviously a tremendous success considering the clarity and quality of the television broadcasts of Winter Olympics 1980. Problems were few and primarily mechanical. The new installation was in place and operational by October 1979. In the figure the system connection is shown. Its first purpose was to transform the Lake Placid telephone facility into an ultrasophisticated communications center capable of handling a wide range of telecommunications services necessary to support the 1980 Olympic events.

Figure 8–14 The Lake Placid Communication System (*Courtesy New York Telephone*)

The lightwave system extends $2\frac{1}{2}$ miles and links the Lake Placid telephone switching office, the Olympic ice arena, and the broadcast center. The broadcast center serves 25 mass-media agencies. The fiber cable, made by Western Electric travels approximately 2 miles on poles (aerial) and the other half-mile underground. Part of the underground cable is buried. The cable consists of 12 glass fibers carrying digital voice signals and television. Six of the fibers carry 288 two-way voice conversations, while two of the fibers transmit the video and associated voice signals. The remaining four fibers serve as backup in the event of failure. An added part of the Olympic system is a microwave installation which serves the entire facility.

A second major effort using fiber optics is the permanent installation linking telephone company central offices between White Plains and East 38th Street Manhattan, New York. (See figure 8–15.) This is an extremely long system (30 miles). It was installed to meet with a growing volume of telephone calls between New York City and Westchester. The New York system will carry both voice and data communications between company central offices at White Plains, Scarsdale, Tuckahoe, and Mount Vernon. This part of the major link will cover 11 miles and have a potential for handling 14,000 calls simultaneously. This part of the major effort is scheduled for completion in 1980.

The second part of this major fiber-optic system will enter New York City from Mount Vernon, pass through city central offices and terminate at the East 38th Street exchange. This second link will cover over 17 miles and is installed in existing cable ducts. The Mount Vernon–mid-Manhattan fiber-optic link is probably the beginning of major usage of fiber optics in the telephone industry.

A BASIC T1 FIBER-OPTIC SYSTEM

Early in 1979, Southern Bell Telephone and Telegraph and Florida Power and Light completed an installation of a T1 system. You may recall that a T1 system consists of 24 voice-frequency channels into a 1.544 megabit per second pulse code modulation (PCM) bipolar digital-bit stream. The system was manufactured by ITT Telecommunications of Raleigh, North Carolina. The unique feature of the system is the use of optical cable to metallic cable conversion in a repeater manhole environment. (See figure 8–16.) The conversion repeaters convert infrared light pulses into electronic pulses for transmission over conventional digital span lines to the central telephone office.

System Description   The computer center consists essentially of three pulse-code modulation (PCM) channel banks, an automatic protection switch (APS) for three service lines with a spare, an optics shelf consisting of optical

Figure 8–15 The New York City and Westchester Fiber Optic Telephone Link (*Courtesy New York Telephone*)

transmitter units (OTU), office power converters (OPC) and optical receiver units (ORU). (See figure 8–17.)

Each PCM channel bank multiplexes 24 voice-frequency (VF) channels into a 1.544-megabit-per-second pulse-code modulation (PCM) bipolar digital-

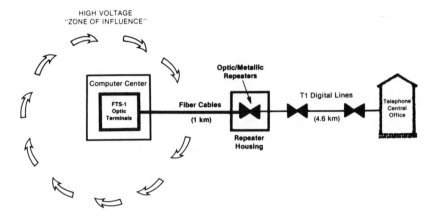

Figure 8–16 A Basic TI System *(Courtesy ITT Telecommunications)*

Figure 8–17 TI System Description *(Courtesy ITT Telecommunications)*

bit stream for transmission. A special test-and-alignment unit plugs into the shelf of the PCM units to measure and adjust the receiver voice-frequency (VF) levels, provide a 1020-hertz test tone for setting transmit voice-frequency levels, and to measure idle channel noise, linearity, distortion, crosstalk, hybrid balance, and office equipment losses.

The automatic protection switch (APS) accepts the 1.544-megabit-per-second signals from the three PCMs. If any of the three optic service lines fails, the APS automatically transfers PCM signals to the spare line. After the fault is cleared, reset to normal is automatic. The optic line shelf consists of optical transmitter units (OTU), office power converters (OPC), and optical receiver units (OPU). The optical transmitter units (OTU) convert the 1.544-megabits-per-second-rate bipolar electrical signal to a 3.088-megabits-per-second optical signal suitable for launching over the fiber-optic line. Wavelength of the launched signal is 850 nanometers. The optical receiver unit (ORU) detects the 3.088-megabits-per-second signal and decodes it back into a 1.544-megabits bipolar signal. The office power converter (OPC) provides the power for operation of the OTU and the ORU.

Three optical cables with five fibers each are routed through underground conduit from the optical line shelf to repeater housings in the manhole. The cables are 7 millimeters in diameter. The fibers are graded-index and have an attenuation of 6 decibels per kilometer.

An optical-to-metallic conversion repeater interfaces the opposite end of the fiber cables to the T1 span line. Each converter has capability of handling three single system repeaters. In this system the two repeaters are also connected together with three fiber cables allowing for expansion. The converter housing is stainless steel and is pressurized to 15 pounds per square inch to purge moisture. Each conversion repeater consists of three sets of conversion units. Each set has an optical-to-metallic regenerator (OMR) that consists of a 3.088-megabit-per-second optical receiver, a bit-rate converter, and a 1.544-megabit-per-second metallic span-line driver. The metallic-to-optic regenerator (MOR) contains a 1.544-megabit-per-second metallic regenerator, a bit-rate converter, and a 3.088-megabit-per-second optical transmitter. The line power converter (LPC) is a dc to dc converter which provides power for the OMR and MOR units. All the units in the conversion repeater are similar in function to the units on the optical line shelf. Metallic cables connect from the conversion repeaters to the T1 digital line in the central office.

A BASIC T3 SYSTEM

A typical T3 communications system spanning a distance of 22 kilometers was scheduled for completion by the Pennsylvania Commonwealth Telephone Company in mid 1979. The system, manufactured by ITT Telecommunications, utilizes four intermediate repeaters between two north-central Pennsylvania communities of Wellsboro and Mansfield. The system carries toll, intertoll, operator, and special service traffic. Optical cables include 10 kilometers of direct buried (plowed) cable, 1.0 kilometer of underground (duct) cable, 3.0 kilometers of aerial cable lashed to existing cable, and 8.0 kilometers of aerial cable lashed to a new messenger strand.

System Description   The modified T3 communications system (see figure 8–18) accepts a 44.736-megabit-per-second electrical signal from a digital radio or a multiplexer switch and converts it to an optical signal. The optical signal is launched into fiber-optic cables and transmitted to a receiver at far-end optical terminal. Sufficient optical repeaters are placed into the link between the two terminals. The receiver detects the optical signal, converts it to an electronic level, and forwards it to the receive side of the far-end terminal where it is converted to T3-level signals and routed to radio or multiplexer.

In the lower part of the figure 8–18, the metallic equivalent of a T1 communication system is compared to the optical system. Electrical signals from digital radio or a digital switch are fed to repeater and transmitted over metallic lines to repeaters at the far-end central office and routed to radio channels. A typical T3 system conversion then consists of a pair of office terminal repeaters and their housings and the fiber-optic cables with connections.

A T3 office terminal is pictured in figure 8–19. Internally, the terminal consists of an optical transmitter, an optical receiver, a looping unit, scrambler and

Figure 8–18   A Basic T3 Communication System (*Courtesy ITT Telecommunications*)

Figure 8–19 A T3 Office Terminal (*Courtesy ITT Telecommunications*)

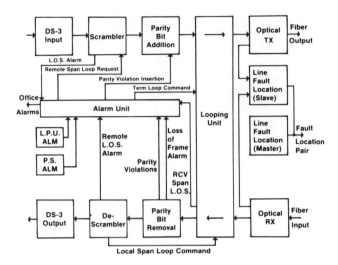

Figure 8–20 Functional Diagram of T3 Office Terminal (*Courtesy ITT Telecommunications*)

descrambler, and an alarm system. See figure 8–20 for a line drawing functional block diagram of the T3 office terminal.

A T3-level signal is routed to and from the T3 terminal. The signal is scrambled to randomize the input, which improves repeater and receiver clock extraction and removes jitters. Parity bits are added to allow bit-error-rate (BER) monitoring and the transmission of system control functions. The 44.736-megabit-per-second signal (now a 47.367-megabit-per-second signal) is routed through the looping unit and to the optical transmitter for transmission. The looping unit allows out-of-service diagnostic testing from either end of the transmission link. One operator at one end of the system can test all units from the input to the output except for the optical receiver and transmitter.

The optical transmitter modulates an injection laser diode (ILD) at a near-infrared wavelength of 850 nanometers. Modulated light is launched into fiber cables and transmitted toward the far-end receiver. The modulated light moves through the fiber cable (through repeaters) and to the optical receiver in the far-end office terminal. The optical receiver employs an avalanche photo-diode (APD) as a detector. The signal is amplified. Clock-timing information is extracted and the signal data stream is regenerated. Parity bits added at the input of the opposite-end office terminal are removed and signals are descrambled.

T3-level signals are then routed to radio or multiplexer. All signals along within the office terminal are monitored in the alarm unit and line-fault information is collected between the optical transmitter and receiver.

Cable Routing Layout   The illustration in figure 8–21 is a line drawing of the fiber-optic cable and its metallic counterpart. Note that the expanse between metallic splices is typically 2 kilometers while the distance between optical cable splices is 1 kilometer. The practical loss allocations, along with a reasonable

Figure 8–21   T3 System Cable Routing (*Courtesy ITT Telecommunications*)

safety margin, result in a 5-kilometer expanse between the optics terminal shelf in the T3 office terminal and the repeater housing. Fiber cables are spliced in the field using a flame-fusion process. A fiber position fixture and a microscope are utilized to control alignment of fiber ends. Splice losses are around 0.5 to 1.0 decibel. Considering that the splicing takes place in the field, in varying kinds of weather, the results are more than satisfactory.

Optical Repeater Housing   Repeater housings like the one in figure 8–22 are designed to be installed on poles or in manholes. When pole-mounted, a sun shield is used to eliminate heat rise that would reduce laser life. The housing contains the same optical receivers and transmitters as those in the office terminal. The case is pressurized. It has an internal electronic card cage designed to hold two bidirectional optical repeaters, power supplies, and repeater performance monitoring and reporting systems. The housing is equipped with metallic stubs for power leads, order wire, and system-monitoring wire pairs. Load coils and lightning protection for the copper pairs are also provided. Optical cables are installed in the field and with an epoxy dam. Power for the repeaters is ± 130 volts direct current to provide a 140-milliampere loop current. Current is routed to the repeater using a 22- or 24-gauge copper pair. At the present time, optical span line electronics cost about the same per voice-channel mile as the metallic versions. Longer length and systems with large numbers of channels tend to make the optical solution a better purchase because of the cable cost savings.

Figure 8–22   T3 System Repeater *(Courtesy ITT Telecommunications)*

SYSTEM COUPLING USING MULTIPLE-ACCESS COUPLERS

The purpose of an optical coupler is to distribute the optical signal into multiple branches of a data transmission system. Couplers (data buses) are usually star type, directional, or are specially made for a specific system. Input signals from numerous branches are mixed together and distributed. Multiple data is outputted by the receiver in a serial fashion. The concepts used are a mirrored surface, which reflects all of the inputs to each of the output channels and a ratio-splitting method using fiber bundles.

Star Couplers Figure 8–23 shows the two basic star couplers, the transmission and the reflection star. These couplers are linear mixing devices, that is, input light signals are mixed together and then divided equally among the output parts. This being the case, the couplers can combine various signals together by a multiplexer or split one signal into various parts. The coupler may also insert light power into or out of a fiber-optic link. Both types are composed of a set of input fibers, a set of output fibers, and a mixing region. The difference between the two star couplers is obvious.

In general, the reflection star is more versatile than the transmitting star coupler because the relative number of input and output parts may be selected or varied after the device has been constructed, while the number of transmitting star inputs and outputs are fixed by initial design and fabrication. However, the reflecting star coupler is less efficient than the transmission star, since a portion of the light fed back into the coupler is injected back into the input fibers. Reflection and transmission star couplers are used where efficiency must be considered.

Figure 8–23 Star Coupler Types (*Courtesy ITT Cannon Electro-Optical Products Division*)

Directional Couplers Directional couplers, also called Tee couplers can be used for signal insertion, monitoring branching, wavelength duplexing, or mixing in multiple-access, bidirectional, point-to-point, or data-bus systems. Optoelectronic directional couplers have a PIN diode within the coupler at a tapoff port. (See figure 8–24.)

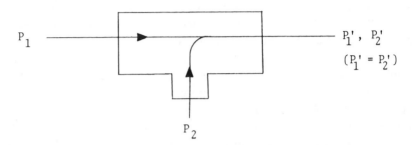

Figure 8–24 Directional Coupler (*Courtesy ITT Cannon Electro-Optical Products Division*)

Systems Using Couplers   With the coming of age of low-loss single-fiber tapoff, mixing, branching, and wavelength multiplexing, users of couplers can intercommunicate, link to a central terminal, or sample data streams along a single-optical-fiber data bus. (See figure 8–25.) Multiple-access couplers, when configured with appropriate light sources, fibers, and detectors allow the user to take advantage of the desirable features of optical fibers. These features, as out-lined in chapter 7, include wide bandwidth, low loss, immunity from radio fre-quency (rf) and electromagnetic pulse interference (EMI) along with small cable size, high transmission security, negligible crosstalk, ground isolation, and short-circuit protection.

In figure 8–25(a), an application using optical and optoelectronic couplers is illustrated. A in the illustration is a directional coupler being used as a laser-stabilizer feedback-loop element. This coupler has a PIN diode in its tapoff port. B1 in the illustration is an optical directional coupler being used as a single-fiber information-drop element to a receiver. B2 is also a directional coupler used as a single-fiber insertion coupler from a transmitter. Directional couplers may be compatible to plastic-clad silica fibers or glass-on-glass graded- or step-index fibers.

In figure 8–25(b), bidirectional couplers (C) are being used to place a duplex communication system on a single optical fiber. Two-color multiplexing provides required channel isolation. The bidirectional coupler is compatible to glass-on-glass fibers.

In figure 8–25(c), both transmission and reflection star couplers are used in a complex mixing system. The transmission star coupler's input parts are optically isolated from each other. The reflection star ports are all coupled. Star couplers are compatible to plastic-clad silica fibers.

While all couplers can be used with any fiber, power loss due to the mismatch of fibers may not be tolerable. Couplers can add that extra freedom necessary in unique system design.

Figure 8–25 System Using Couplers (*Courtesy ITT Electro-Optical Products Division*)

Data Buses The major use for fiber-optic multiple-access couplers is in multiuser, multiservice fiber-optic buses. Within a typical bus are terminals, nodes, multiplexers, and links that comprise the network. In addition, the bus is separated into access and center user networks. Local users are grouped and interact in access networks. Center users are interconnecting. Services that can be handled on a fiber-optic bus include digital data, voice, video, and informational retrieval. These services can be segregated by frequency-division multiplexing. High-bandwidth fiber enables many channels to be multiplexed on a single-fiber bus.

Loop-Bus and Star-Bus Configurations   Both star and directional couplers can be utilized to develop a data-bus configuration. Figure 8–26 represents typical star-bus configurations, while figure 8–27 represents a loop-bus configuration. The star bus requires a centralized selection process implemented by a branching

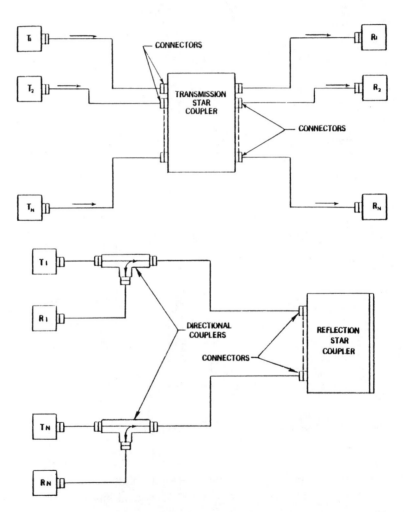

Figure 8–26   Star Bus Configurations (*Courtesy ITT Electro-Optical Products Division*)

element. The loop networks pass all information through each node and require drop-and-insert elements to provide access to receiver and transmitter terminals. In figure 8–26, both transmission and reflection star couplers are illustrated. The reflection-star configuration requires only half the cable used by the transmission-star configuration. However, power is sacrificed and an additional directional coupler is required for each channel. An alternative reflection-star coupler may have a single part for each transmitter and receiver. Figure 8–27 illustrates a loop-bus configuration consisting of three transmitter and three receiver directional couplers. Fibers are pigtailed. The choice between the star and loop configurations depends entirely on system requirements. Advantages of the loop configuration are expandability, ease of synchronization, and low cable cost. However, the loop configuration has accumulated insertion loss as a data stream passes drop points. Advantages of the star bus are the availability of a multiple of repeaterless terminals and a great savings in cable cost, therefore lower prices to consumers. It can also be easily expanded. The star bus has disadvantages in that synchronization is difficult because of the varying propagation delays from the user terminals.

Figure 8–27 Loop Bus Configuration (*Courtesy ITT Electro-Optical Products Division*)

Data-Bus Multiplexing After the multiple-access couplers have been selected, a method for multiplexing several messages onto a single channel and a method to control and route data traffic are required. Multiplexing techniques in conventional wideband radio frequency (rf) networks are characterized by modulating carrier waveforms in the time domain. Two common methods of carrier modulation are time-division multiplexing (TDM) and frequency-division multiplexing (FDM).

INTERFACING DATA PROCESSING EQUIPMENT

One of the more typical and usable fiber-optic links in today's market is the Canoga Data Systems CRS–100 Data Link (see figure 8–28). The data link operates in a data range from dc to 56 kilobits per second (kbps) over a distance of 3000 feet with an operational bit error rate of 10^{-9} or better. The unit is powered by a standard ac wall outlet plug (the stand-alone unit) or may be mounted in a powered rack such as that in the figure 8–28. The link's transmitter and receiver operate at an input/output wavelength of 820 $\pm$15 nanometers (nm). The light source for the transmitter is an infrared LED, while the detector for the receiver is a PIN diode. The link has all the advantages found with fiber-optic systems. These are elimination of electromagnetic interference (EMI), equipment electrical isolation, and data transmission security. Some of its applications follow.

Figure 8–28 The CRS 100 Fiber-Optic Digital Data Link (*Courtesy Canoga Data Systems*)

Simplex Application Probably the simplest fiber-optic data link is a one-directional link between a CPU and a single card reader, plotter, printer, and paper-tape reader. The simplex version of the CRS 100 data link uses the CRS 101 transmitter and the CRS 102 receiver. The CRS 100 operates in transmit and receive modes simultaneously or can operate in half duplex mode (receive or transmit). (See Figure 8–29.)

Figure 8–29 Digital Simplex Application (*Courtesy Canoga Data Systems*)

Duplex Application When two-way communication is required, the duplex CRS 100 link is utilized. A single terminal setup is shown on the top of figure 8–30. The same communication link can serve several terminals if they are located within 50 feet of a common point and messages are short and/or infrequent enough that satisfactory operation is possible with only one terminal transmitting at one time. This can be accomplished in two different ways. The first is a straight polling configuration. See the center setup in figure 8–30. Terminals are connected in a series/parallel arrangement and one terminal is connected to the data link. The second method is the use of a modern sharing unit. See the lower setup in figure 8–30. Modem (link) sharing units typically have four or eight terminals and may be cascaded as shown in the lower illustration. All the setups shown in these illustrations have one thing in common, the transmission lines between major units are optical fibers.

Figure 8–30 Digital Duplex Applications (*Courtesy Canoga Data Systems*)

IN-SITU TESTING

The requirement for testing does not stop with the laboratory or the manufacturer's test line. Once installed in the field, faults must be located and repaired and parts replaced. One of the most prevalent types of equipment utilized in field testing is the optical time domain reflectometer (OTDR). A typical OTDR is the Siecor Model 38, shown in figure 8–31. The Model 38 has extensive capabilities. It is used to determine the length of optical fibers or cables, locate fiber fractures and anomalies, evaluate splice or connector continuity, measure cable attenuation, measure splice and connector attenuation, and assess fiber homogeneity. Furthermore, the Siecor Model 38 OTDR can be used to determine cable properties before and after installation or during pulling/burying operations, along with the monitoring of field splicing and connecting processes.

The Model 38 OTDR has a built-in alignment mechanism that allows it to accommodate a variety of fiber geometries and terminated cables. The OTDR utilizes the effects of backscattering and reflections. Measurements are based on transit time and signal amplitude. The OTDR is portable and has only a few connections and controls, which make its use fundamental.

Figure 8–31 The Siecor Model 38 Optical Time Domain Reflectometer (OTDR) (*Courtesy Siecor Optical Cables, Inc.*)

A typical setup for attenuation and length measurements using the OTDR is shown in figure 8–32, along with an electrical/optical block diagram. Note that the OTDR requires access to only one end of the optical cable. Note also that it allows accurate measurement of attenuation in extremely long runs of cable. Typical ranges of cable runs today are 3 to 6 kilometers, with connectors at 1-kilometer positions. Typical splice losses are 0.3 or 0.4 decibels, while cable attenuation is 3 to 5 decibels per kilometer.

In figure 8–32, a laser diode is driven by a laser pulser that receives its power from a dc/dc converter. A laser switch sets the test in operation. The injection laser diode emits light pulses in the near-infrared. The pulse passes through a collimating lens and a 3-decibel beamsplitter and is launched into the fiber under test through a second lens. The reflections from the near end and from the far end are diverted by the beamsplitter and received by the avalanche photodiode.

Illustration of OTDR ranges in a typical
cable system

Figure 8–32 Typical Cable Length and Attenuation Test Setup (*Courtesy Siecor Optical Cables, Inc.*)

The APD signal is amplified and sent to an external oscilloscope through a signal output jack for visual display. The fiber length is calculated from the difference in arrival time of the near- and far-end reflections. Attenuation is computed from the slope of the backscattering curve shown on the oscilloscope. Details on oscilloscope operation are found in vendor catalogues. Measurement techniques using the Siecor Optical Time Domain Reflectometer (OTDR) are established in corporation application notes and are provided with the equipment.

9

LASERS

GENERAL

As described in chapter 1, laser technology is the science that deals with the generation of coherent light in small but powerful beams. The word *laser* is an acronym originating from the initials of *l*ight *a*mplification by *s*timulated *e*mission of *r*adiation. The operating frequency range of the laser is in the optical frequency-spectrum from 0.005 microns to around 4000 microns. (A micron is equal to 10^{-6} meter.)

There are four basic laser types, crystal lasers, gas lasers, liquid lasers, and semiconductor lasers. Crystal lasers use a crystal such as the ruby for an active medium. Gas lasers use gases in combination, such as helium and neon, as an active medium. Liquid lasers use organic dyes such as rhodamine as an active medium. The semiconductor laser is made from semiconductor material such as gallium arsenide. The semiconductor laser was covered in detail in chapter 6 under "Injection Laser Diodes."

All lasers have some common characteristics. These are an active material to convert energy into laser light, a pumping source to provide power or energy, optics to direct the beam repeatedly through the active material to become amplified, and optics to direct the beam into a narrow cone of divergence. Other characteristics common to all lasers are a feedback mechanism to provide continuous operation and an output coupler to transmit power out of the laser.

Laser technology is extremely diverse and complex. This section of the book was added to give a general overview of lasers and is not intended to provide detailed physics and description of the laser field, which would take volumes.

LASER LIGHT PROPERTIES

Laser light has unique properties. A laser beam is one color or monochromatic. If the light is one color then it is also one wavelength, and is therefore coherent. An example of incoherent light is that of an ordinary light bulb, which emits white light. White light is incoherent because it has all visible colors and therefore all visible wavelengths. Figure 9–1 shows some laser light properties. Light such as the light bulb is panchromatic (has all visible wavelengths) and cannot be concentrated or controlled. Laser light can be controlled because it is monochromatic (one color), one frequency. The beams are in-step (in-phase) with each other.

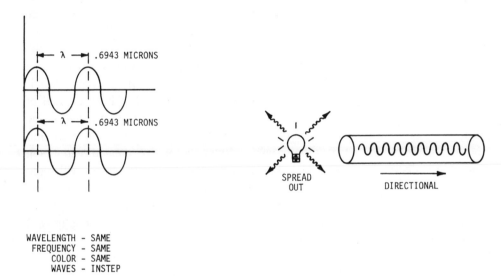

WAVELENGTH - SAME
FREQUENCY - SAME
COLOR - SAME
WAVES - INSTEP

Figure 9–1 Laser Light Properties (*Mike Anaya*)

The light in a laser beam is extremely directional. It is focused by a lens into a narrow cone and projected in one direction. The direction of a laser beam is the result of optics at the ends or sides of an active medium. The radiance of the beam can be compared with the intensity of the sun. The laser, when focused to a fine, hairlike beam, can concentrate all of its power into the beam. Hence the directionality of the beam has a bearing on its intensity. If the beam were allowed to spread out it would lose its power. Confinement of large amounts of power into a narrow slit allows the laser to be used to drill holes in and cut metal, as well as welding applications.

Lasing Action   We have discussed the fact that incidence radiation could change the energy state of an atom from the ground state to an excited state. We also know that if left alone for a small time period in the excited state that the atom's energy will decay and fall to the ground state and produce a photon. For amplification by stimulated emission to happen, there must be a chance for the created photon to meet other atoms in the excited state and a population inversion must be created. That is, there must be more atoms in the excited state than in the ground state. This situation can happen by a process called *pumping.* In this process, energy is pumped into the active medium to raise atoms to an excited state. This must be done very quickly, because the atoms in the excited state may spontaneously decay to the ground state before amplification takes place. To get quick response action a high-intensity lamp is required with a very large power supply. With a low-intensity lamp only incoherent light will be produced. In a three-energy-level band such as a ruby laser, the pumping action raises the atoms to the excited state which absorbs a large range of wavelengths. These atoms in turn spontaneously drop to an intermediate energy state. Heat is dissipated. However, during the process, a population inversion has been created at the intermediate level. Now that the population inversion has been created, lasing action can take place. Soon atoms drop to the ground state and in doing so emit photons. These photons strike other atoms and two photons are created, and so on. Thus stimulated emission has started lasing action and soon the beam will build up to a large strength. When the beam strikes the mirror surfaces of the active medium, it is reflected back and continued lasing takes place. Because of this reflection and continual lasing action an intense beam is built up. Figure 9–2 shows a schematic of lasing action.

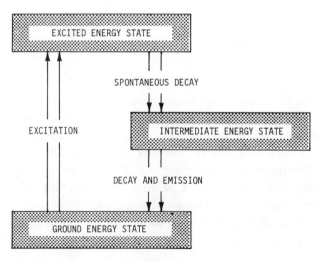

Figure 9–2 Lasing Action (*Mike Anaya*)

LASER CLASSIFICATIONS

Lasers are classified in three ways:
1. Their active mediums
    a. Solid
    b. Gas
    c. Liquid
    d. Semiconductor
2. Their wavelength
3. Their output time characteristics
    a. Mode-locked pulse
    b. Q-switched pulse
    c. Normal-mode pulse
    d. Continuous wave (CW)

Active Medium Classification   The solid-crystal ruby laser was the first one developed. As discussed in chapter 1, Theodore H. Maiman, a scientist at Hughes Aircraft Company, created the first laser from a ruby crystal. The solid lasers in use today are characterized by their active medium (host material), the dopant they use for lasing ions, and their output. The major solid lasers are the ruby, Nd:YAG, and the Nd-doped glass. The medium for the ruby laser is aluminum oxide and its active ions are chromium. Chromium is also known as a transition metal. The Nd:YAG laser uses a rare-earth element called neodymium. The neodymium (Nd) is dissolved in the host crystal yttrium aluminum garnet (YAG). The third solid laser is the Nd-doped glass laser. Glass differs from other solid lasers in that it is not a crystalline material. It is, however, cheap to produce in any shape and can be more efficient than the ruby. The host glass is employed with an active ion usually neodymium (Nd). Another atom (uranium), part of a group called *actinides,* is used as an active ion.

Active mediums of gas are classified as neutral-atom, ion, and molecular. The neutral-atom gas lasers have active-medium atoms which have an equal number of charged protons and electrons. A type of neutral-atom laser is the helium-neon (HeNe) laser. An ion gas laser has an active medium of ionized atoms (have one or more electrons moved from their shells) and has a positive charge. A type of ion gas laser is the singly iodized argon Ar +.

A molecular gas laser has an active medium which has two or more atoms either of the same or of different atomic species. A type of molecular gas laser is carbon dioxide (CO_2) laser. Liquid lasers typically use organic dye dissolved in alcohol or water and are usually pumped with an argon laser. A type of liquid laser is one using rhodamine 6G dye. Semiconductor lasers are described in detail in chapter 6.

Wavelength Classification It is important that wavelength be considered when classifying lasers. It was a giant step that moved the laser out from the classification of the maser. The optical range of wavelengths, as stated many times in this book, have three major areas: ultraviolet, visible, and infrared. Wavelengths in the ultraviolet spectrum range from 0.005 to 0.3900 microns. Wavelengths in the visible spectrum range from 0.3900 to 0.7500 microns. Wavelengths in the infrared spectrum range from 0.7500 to 4000 microns. Table 9–1 lists typical lasers along with their wavelengths.

TABLE 9–1 Typical Lasers and Their Wavelengths

Laser Type	Wavelength (in microns)	
Hydrogen cyanide (HCN)	773	
Water vapor (H_2O)	118.6	
Carbon dioxide (CO_2)	10.6	
Hydrogen fluoride (HF)	3.1	Infrared
Yttrium aluminum garnet (Nd:YAG)	1.064	
Glass (Nd:glass)	1.060	
Gallium arsenide (GaAs)	0.9100	
Krypton (K)	0.7525	
Ruby (Al_2O_3)	0.6943	
Helium-neon (HeNe)	0.6328	
Rhodamine (6G)	0.5900	Visible
Argon (Ar)	0.5145	
Argon	0.4880	
Helium-cadmium (HeCd)	0.4416	
Molecular xenon (Xe_2)	0.1700	Ultraviolet
Hydrogen (H_2)	0.1600	

Output Time Characteristics Output time characteristics fall into categories of time duration and power. The first of the characteristics (or *modes of operation* as they are called) is the mode-locked pulse. Mode locking is a process in which longitudinal modes circulating in a laser cavity have a fixed difference on separation. The effect is used in solid lasers by inserting a cell with a special organic dye. Mode-locked lasers operate in ultrashort pulses of about 10 picoseconds at 10^4 to 10^{12} watts of power.

Normal pulsed lasers, such as the first ruby laser, store energy in a capacitor bank that discharges into a flashlamp, which in turn radiates light into the ruby. This causes chromium atoms to move to their excited states. Single pulses are emitted with uneven spikes at a few milliseconds in duration at 10^3 to 10^5 watts of power. The pulses are not equal in amplitude.

Q-switched lasers replace the 100 percent reflecting mirror on one end of the ruby rod with a high-speed rotating prism. The rotating prism develops a greater population inversion of atoms within the ruby. Pulses are cleaner than those of the normal pulse laser. Q-switched laser pulses are about 25 nanoseconds in duration at 10^6 to 10^9 watts of power. A continuously pumped repetitively Q-switched laser provides repetitive pulses of close to 50 nanoseconds in duration with up to 50 kilowatts of power.

SOME LASER TECHNICALITIES

Energy Levels Each atom has its own light spectrum. That is, each material has a set of wavelengths called *spectral lines* that are representative of the material. Each atom also has its own set of energy levels, which is a result of transition between two of the energy levels. The energy of the levels is usually expressed in electron volts (eV), but may also be discussed in terms of joules or ergs. Some comparisons of these units are listed below.

$$1 \text{ joule} = 10^7 \text{ jerg}$$
$$1 \text{ volt} = 1 \text{ joule/coulomb}$$
$$1 \text{ electron volt} = 1.602 \times 10^{-19} \text{ joule}$$
$$1 \text{ electron volt} = 1.602 \times 10^{-12} \text{ jerg}$$

Measurement of these spectral lines is accomplished using a device called a *grating spectroscope*. The spectroscope has a narrow slit through which light enters. The light is partially collimated through an aperture and may be monitored on a wavelength scale. The wavelength scale is superimposed on a pattern of colored bright and dark lines representing the color spectrum of the light. The wavelength is measured in angstroms.

A Three-Energy-State Laser The ruby is usually a three-energy-state atom. The top shaded area in figure 9–2 is the excited energy state. Dozens of energy levels are crowded at this energy state to form an energy band. The second energy state down is an intermediate energy state. The lowest energy state is ground. Pumping raises the chromium atoms from ground energy state to excited energy state. This is called *stimulated absorption*. The ions then decay and fall to a lower or intermediate energy state. When the population of ions in the intermediate state is greater than the ground state, population inversion is present. That is, there are now more excited atoms in intermediate than ground state. When the population inversion is present, laser action (lasing) can take place. In time, the excited chromium atoms will fall to the ground energy state. When they do so, photons are emitted. A photon you may recall is a packet of light caused by radiant energy. The frequency of the energy determines the strength of the photons. Higher frequencies cause greater-strength photons.

Now let us consider the term *stimulated emission*. Chromium atoms are pumped to an excited energy state, then go through a spontaneous decay to an intermediate energy state where an accumulation of excited atoms creates a population inversion. As time progresses, the excited atoms fall from the intermediate energy state to the ground state. When they do, photons are emitted. As the photons are emitted they strike excited atoms and two other identical photons will emerge. These two photons in turn strike other excited atoms and stimulated emission occurs. The more often this happens the greater the intensity of the light beam. The ruby laser just discussed produces laser light in pulses. It is possible for the ruby laser to produce continuous pulses. If this is the case, energy must be pumped continuously into the optical cavity to maintain population inversion. A problem of heating results but can be resolved using a four-energy-state laser.

Attenuation versus Amplification   In figure 9–3, the difference between attenuation and amplification is explained simply with two separate mediums. When a light signal is directed or pumped into a medium of glass, part of the light signal is absorbed by the glass and part allowed to continue through the glass. This phenomena is called the *exponential law of absorption*. The law involves the intensity of the transmitted light signal, the thickness of the material, and the intensity of the incident light. In the ruby medium a light signal is directed or pumped into the ruby medium causing population inversion and stimulated emission rather than absorption.

Figure 9–3   Attenuation versus Amplification (*Mike Anaya*)

Coherence   Coherence of a laser beam must be explained by way of phase relationship. Spontaneous emission may move in any direction. This is not the case of stimulated emission. Stimulated emission has the same phase as the applied or incident radiation and indeed is the same shape. Furthermore, the radiation is moving in the same direction. When the incident and stimulated-emission wavelengths are in phase along the beam, this is called *temporal coherence*. When the same relationship occurs across the beam, it is called *spatial coherence*. The importance of coherence with lasers cannot be understated. A coherent beam means that maximum power can be focused in an extremely small area. With an incoherent beam this is not possible. The application of a monochromatic light source would seem imperative in laser applications.

Laser Beam Divergence   The light beam emitting from a laser has an extremely small diameter at its aperture. The beam does not continue at this same diameter but diverges (spreads out) into a larger diameter forming a *cone of divergence*. The divergence angle is usually very small. Gas lasers have the smallest divergence angle, somewhere near 1 milliradian. You may recall from mathematics that a radian is 57.3°. A milliradian is 0.0573°, which equates to 3.44 minutes of arc. Liquid lasers may have a 10 to 20 milliradian divergence angle. When specifying a divergence angle it is often given as a half-angle. Solid lasers have a divergence angle of 5 to 25 milliradians. Semiconductor lasers such as the injection laser diode often have a divergence angle of 500 milliradians. Laser beam divergence and lens focusing are important in laser alignment. Laser alignment is covered in greater detail later on in this chapter.

The Optical Cavity   The amplifying medium of a laser is usually between a pair of reflecting mirrors. These mirrors serve as optics to direct the electromagnetic waves into and out of the active medium. Light waves are reflected back and forth through the cavity and bounce off the mirrors. At the surface of the mirror are incident waves (toward the mirror) and reflected waves (away from the mirror). You must remember that these waves are the same frequency. When they move in opposite directions, a pattern results if they interfere with each other. When they interfere a phenomena exists in that the waves appear to stand still. If two waves of the same frequency and amplitude interfere with each other by moving in opposite directions, the result is a standing wave pattern (see figure 9–4). The standing wave will only happen if the relationship between the wavelength and the length of the cavity are compatible. Lasing operation will only occur for typically two or three frequencies. Only those frequencies or modes of oscillation that satisfy the relationship are compatible. The formula below relates to this compatibility.

$$N = \frac{2n_0L}{\lambda}$$

where   N = number of half-wavelengths or loops that will fit between the mirrors in the optical cavity.

n_0 = refractive index of the active medium.

L = length of the optical cavity

λ = wavelength

Figure 9–4   Standing Waves in an Optical Cavity (*Mike Anaya*)

A Four-Energy-State Laser   The four-energy-state laser resolves the heating problem of the three-level ruby laser. Typical of the four-layer laser system is the Nd:YAG laser. The top shaded area in figure 9–5 is the excited state. Dozens of energy levels are crowded at this energy state to form an energy band. The second state down is an intermediate state to where atoms spontaneously decay. A secondary energy state is a level of spontaneous decay and stimulated emission, while the ground state is at the lower energy state level.

The process of pumping raises the neodymium (Nd) from ground energy state to excited energy state. The ions decay and fall to a lower or intermediate energy state. No radiation takes place between the excited and intermediate states. Lasing action takes place between the intermediate and secondary energy states.

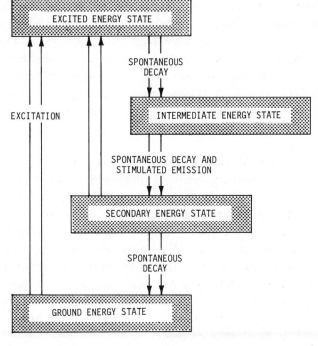

Figure 9–5 A Four Energy State Laser (*Mike Anaya*)

These two center levels have a lower population than the ground state. The secondary has a lower population than the intermediate. The difference, however, between the secondary and intermediate states is still much lower than that between secondary and ground, therefore much less energy is required to maintain a population inversion between the two center levels. Much less energy has to be pumped into the laser cavity and the laser operates continuously instead of in pulses. It must be noted that all four-energy-state lasers do not operate continuously. Likewise not all continuously operating lasers are four-layer.

SOLID LASERS

There are three basic solid lasers: the ruby laser, the Nd:YAG (yttrium aluminum garnet) laser, and the Nd:glass laser. The ruby laser operates in the visible range of the electromagnetic frequency spectrum at 0.6943 microns wavelength. The Nd:YAG and Nd:glass lasers operate in the near-infrared range of electromagnetic frequency spectrum at 1.0640 and 1.0600 microns wavelength, respectively.

The ruby as an active medium consists of aluminum oxide with small amounts of chromium dissolved in it. Rubies used in lasers are formed in a high-temperature crucible by melting the aluminum and chromium and inserting a seed crystal of ruby. A drawing process is used with the seed crystal being pulled at a constant speed while rotating. The ruby acts as the host crystal while the chromium acts as the activator. The Nd:YAG laser uses a host crystal of yttrium aluminum garnet and an activator of neodymium. The Nd:glass laser uses a host material of glass and an activator of neodymium.

Solid-State Laser Operation   As an introduction to solid lasers let us take a cursory look at their operation. A typical solid laser consists of an active medium, an excitation source, a feedback mechanism, and an output coupler. The active medium in the solid laser illustrated in figure 9–6 is a ruby. The typical active medium is a tubular-shaped crystal structure. The ends of the crystal are highly polished and are cut parallel to each other. These ends are coated with a reflective coating. One end of the ruby reflects 100 percent of the light internally. The other end acts as a transmitter, allowing only a small portion of the light to pass outward while reflecting most of the light back into the ruby for stimulated emission. The excitation source is usually a flashtube. The tube is fitted with a trigger which emits a burst of light to start the process.

Let's now go through the operation and consider how it all comes together. The flashtube is energized by the trigger. It produces a high-level burst of light similar to a camera flashbulb. The flash, being adjacent to the ruby cylinder, causes the chromium atoms within the crystalline structure to become excited. Population inversion, stimulated emission, and lasing action take place.

Figure 9–6   Solid Laser Operation (*Mike Anaya*)

When the beam of photons strikes either mirror ends of the ruby, the reflection of the light beam back into the ruby causes an even greater beam intensity because of photon action. The beam is red in color because the laser is amplifying at a wavelength of 6943 angstroms (0.6943 microns). The white light of the flashtube produces many wavelengths; however, only one wavelength can excite the atoms from ground state to excited energy state.

The amplified light is fed partially out of the laser by a transmitting mirror. The transmitting mirror reflects most of the light back into the ruby crystal to attain an ongoing process of stimulated emission and amplification. If too much energy is coupled out of the transmitting mirror the stimulated emission will decay and die. Therefore, the excitation mechanism must constantly pump energy into the laser.

Q-Switched Solid Laser Operation The most powerful of all laser pulses are those developed by the Q-switched laser (see figure 9–7). The phenomenon is called the giant pulse and operates on the premise that the atoms in the active region remain at the excited state much longer than in other lasers (perhaps 2 or 3 thousandths of a second). In the initial moments of operation, the light path between the 100 percent-reflecting mirror and the partially reflecting mirror are blocked so that lasing action cannot take place. You may recall that these mirrors may be external to the active region or may be coated on the regions ends. On the Q-switched solid laser, the 100 percent-reflecting mirror has been replaced by a prism.

The active material is pumped by a high-power light source to ensure that a large population of atoms is in the excited state. Population inversion is ensured

Figure 9–7 Q-Switched Solid Laser Operation (*Mike Anaya*)

and lasing action can take place. All the energy previously stored is released in one giant pulse. The pulse lasts between 10 to 100 millionths of a second. These are short pulses and dependent on the laser.

The prism is rotated at high speed. Lasing can only occur when a population inversion takes place. Stimulated emission amplification far exceeds losses. The rotating prism is called a Q-switch. Lasing action can only take place when this mechanical switch rotates to a point where it is aligned with the partially reflecting mirror on the opposite end of the cavity. The Q-switch can only be rotated at a mechanically controlled speed. Therefore there are limitations to mechanical hookups and switching times. The function of the switch is to allow time for the active medium to be pumped up before the next pulse. The active medium then becomes a storage place for energy between pulses.

Electro-optical Q-switches provide faster switching times by changing optical characteristics with the use of polarized voltage. These are the Pockel's cell and the Kerr cell. Chemical Q-switches are used for extremely short pulses of a few picoseconds.

A SOLID-CRYSTAL LASER

A typical solid-crystal laser is the Nd:YAG MY-series designed by William D. Fountain and manufactured by the Molectron Corporation, Sunnyvale, California. We shall take a longer look at the Fountain design later on, but let us first consider the development of the Nd:YAG laser and its growth patterns.

Origins of the Nd:YAG Laser You may recall that the first solid laser was a ruby laser developed in 1960 by T. H. Maiman, a scientist with Hughes Aircraft Company. It was through Maiman's research that other scientists were encouraged to look for solid materials that could be used as laser material. One of these solids was from a group of rare-earth materials. The rare-earth material neodymium was dissolved in yttrium aluminum garnet and was found to easily operate within a laser. In 1964, the Nd:YAG laser was created. The Nd:YAG laser operates in pulsed or continuous-wave (CW) modes. Early versions produced a pulse 100 microseconds in duration. These pulses were a group of much smaller pulses. Better control of the laser was made by addition of the Q-switch. The Q-switch resulted in shorter pulses (10 to 30 nanoseconds duration) and with extremely high peak power. Pulse lasers today easily produce megawatts of peak power output.

The first Nd:YAG lasers produced an output beam with a bell-shaped (Gaussian) profile and low divergence (see figure 9–8a). The Gaussian TEM_{00} or

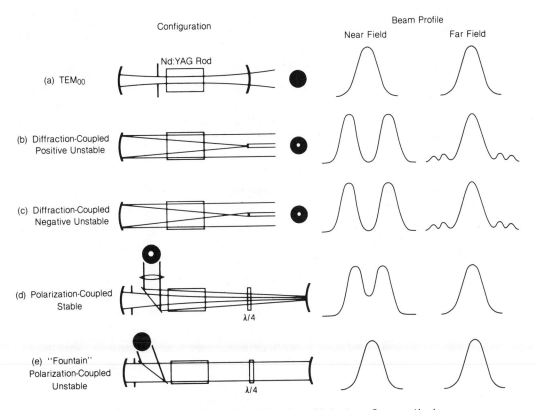

Figure 9–8 Nd:YAG Cavity Design (*Courtesy Molectron Corporation*)

fundamental mode beam is called the optimum beam profile. A great deal of effort is spent in laser design to produce the Gaussian beam. Unfortunately, the TEM_{00} cavity is limited in the amount of energy that can be extracted from a YAG rod. The only way to achieve high energy and low divergence with this cavity is to add amplifier stages.

A diffraction-coupled positive unstable resonator was introduced in 1976 as an alternative to the multiple amplifier TEM_{00} system. (See figure 9–8b.) This cavity is known as diffraction-coupled because the beam exits and diffracts around a small total reflecting output mirror. The term positive unstable refers to the positive curvature of the output mirror. This cavity produces high pulse energy and low divergence in a beam that has a planar or uniphase wavefront. The pulse duration depends on both cavity length and pump energy. The beam profile is that of a doughnut in the near field. Fresnel rings also appear in the intermediate field. These rings are caused by diffraction from the edges of the Nd:YAG rod and fade away in the far field. One of the characteristics of this design is that the doughnut does not persist. Because the beam is diffracted sharply around the output mirror the beam profile constantly changes as a function of distance from the laser. In the intermediate field, a small central spot appears which has extremely high-power density, up to four times that of the doughnut. This spot is potentially dangerous to optical components. The spot causes a discontinuity in the refractive index of transmitting materials and leads to self-focusing and material damage. In the far field (10 meters from the laser) the beam profile has only 60 to 80 percent of the energy of the central lobe.

Figure 9–8c shows the optical schematic of the negative unstable cavity. It is termed negative because of the concave output mirror. The optical and mechanical tolerances of this design are less stringent than those of the positive versions. As a result, it is well suited for lasers on military equipment. For mirror curvatures identical to its positive counterpart, the cavity is twice as long and produces a longer pulse. Otherwise the output is identical to the positive version.

Late in 1977, a stable resonator using a polarization output coupling was introduced. (See figure 9–8d.) This design employs two factors different from previous design. Firstly, polarization techniques couple the beam from the cavity. Secondly, it uses the oscillator rod to increase the output energy. The basic resonator is a TEM_{00} configuration formed by a concave rear mirror and partially reflective front surface of a quarter-wave plate. The aperture restricts the oscillator to the center of the rod and the resulting beam polarized horizontally is low in energy. This beam exits the cavity through the quarter-wave plate $\lambda/4$, which converts it to circular polarization. When reflected off the convex mirror and passed back through the $\lambda/4$ plate, the expanded and diverging beam is polarized vertically, then amplified in the YAG rod and coupled out by the polarizer. A lens then recollimates the beam. Although the oscillator output is the same as the TEM_{00}, the beam exiting the polarizer is not. The oscillator has depleted the gain at the rod center so that the near-field beam has a prominent dip in the center. Fresnel rings are also present. All of the horizontally polarized oscillator beam is converted to vertical polarization for coupling out of the cavity. At high pump energies, unwanted feedback upsets cavity oscillations and limits the energy and quality of the output beam.

All these cavity configurations proved to have good and poor points. In 1978 a beam configuration was designed by William D. Fountain which has no modulation on the beam and its intensity decreases away from the beam center. Beam profiles were close to Gaussian (bell) shaped. The beam maintains high pulse energy and low divergence. (See figure 9–8e.)

The Fountain Design   Mirrors M1 and M2 in figure 9–9 form the optical cavity. In the region to the left of the intracavity polarizer, the beam is polarized horizontally. In passing the quarter-wave plate ($\lambda/4$), the beam becomes elliptically polarized. After reflecting off mirror M2 and passing back through the wave plate, the beam is polarized 75 percent in the vertical plane and 25 percent in the horizontal plane. The vertical component is coupled out of the cavity by the polarizer, while the horizontal component remains in the cavity as feedback. The horizontal component of the beam that remains in the cavity for feedback must pass back through an aperture before reaching mirror M1. The aperture acts as a loss mechanism, which allows energy to build up. Energy lost at the aperture is small.

Figure 9–9 Nd:YAG Laser Operation (*Courtesy Molectron Corporation*)

The Fountain oscillator magnifies the concentric rings produced in previous designs so that the rings are eclipsed by the aperture leaving only a uniform central spot. After the beam exits the cavity, it is incident on a second polarizer which is used to improve the polarization ratio before amplification. Two turning mirrors direct the beam along the rear beam line of the laser.

In an oscillator-alone configuration a beam-reducing telescope is used to increase the power density on the harmonic crystals. A complementary beam-expanding telescope on the output brings the beam back to its original size and also permits adjustment of the output divergence.

In an oscillator/amplifier configuration the amplifier stage replaces the beam-reducing telescope, which is not needed. A negative lens is also added before the amplifier to expand the oscillator beam and to compensate for the positive thermal lensing of the amplifier rod.

In both configurations the beam enters a sealed harmonics box where second, third, and fourth harmonics are generated. These are SHG, THG, and FHG respectively. Each harmonic crystal is hermetically sealed in a cell and kept at a stable temperature. The crystals are angle-tuned with thumbwheels accessible from outside the laser. Before the crystals, a half-wave plate ($\lambda/2$) rotates the polarization of the beam to an axis appropriate to the crystal used. Following the crystals, beamsplitter pairs separate sequentially the shortest wavelength from the combined beam. Following the beamsplitters, half-wave plates ($\lambda/2$) may be inserted to produce any desired polarization axis. The beamsplitters and wave plates serve to permit any combination of these beams to be directed out of the three exit parts of the laser. The center beam line pumps a dye laser directly. The rear beam is used to mix with the dye laser outputs. The front beam is reserved for the residual fundamental beam.

The MY Series Laser   The MY series laser provides high efficiency and high reliability with an output beam that is near Gaussian (bell-shaped). The output pulse duration is 15 to 30 nanoseconds with excellent pulse-to-pulse energy and peak-power stability. The laser has a programmed Q-switched power supply. The oscillator and the amplifier both use high-efficiency power supplies which have a switching rate of 12 kilohertz. Both supplies are current and voltage regulated. The oscillator power supply has a maximum power delivery of 600 watts. The amplifier power supply has a maximum power delivery of 1200 watts. A closed-cycle cooling system of deionized water is used to cool the laser heads and the beam pump. The system is filled with an alcohol-water mixture for shipment. Tap water is required for operation at 2 liters a minute.

A control panel controls power and indicates that the beam is on. A service panel contains voltage controls for the oscillator and amplifier, repetition rate controls, and Q-switch controls. The pump cavity contains the Nd:YAG rod and flashlamps. The Q-switch is a Pockel's cell (an electro-optical device). The Q-switch power supply applies high voltage to the switch crystal only during a flashlamp pulse. A laser safety and interlock system mounts on the outside of a laboratory door to indicate the status of the laser inside. A green light on the interlock indicates that the laser is off. Yellow of course means caution and entry by permission only. A red light means that the laser is on. A door interlock will close the beam shutters on the laser if the laboratory door is accidently opened.

The beam is 8 millimeters in diameter. Pulses are selective at 1064, 532, 355, and 266 nanometers (visible and infrared). Pulse durations are 15 nanoseconds or 15 to 30 nanoseconds variable. Maximum average output power is 20 watts. The laser is a class IV laser product and can cause damage to the skin and/or the eyes.

Design Characteristics of Pulsed Nd:YAG Lasers   The optical cavity must have mirrors that are perfectly aligned. Initial alignments must be within 25 microradians and must be stable for long-term operation. Focal length of the YAG rod varies inversely with the average power delivered to the rod from the flashlamps. If the focal length deviates from design parameters, performance deteriorates in divergence then pulse energy. (See Figure 9–10.)

Figure 9–10   An Nd:YAG Laser (*Courtesy Molectron Corporation*)

Q-switches have crystals with a limited lifetime at high voltage due to ion migration from electrode into crystal, which reduces transmission at 1064 nanometers. Programmed Q-switch power supplies can turn off the power after each pulse.

Optics must be kept extremely clean. High-energy pulses place a severe strain on optical components. The smallest amount of dirt on a mirror surface can cause coating damage. Most lasers use a sealed beam to protect the optics from dirt.

If one mode of the laser is active, the pulse is smooth. If more than one mode is active, beating between modes produces a modulation on the temporal profile (phase relationship at two points in time). Short cavities produce shorter pulses than long cavities, but the transit time is less so the beats are closer together. Nd:YAG rod size is dependent on whether the rod is used as an oscillator or an amplifier. Oscillator rods should be fine quality. Amplifier rods need not be as high quality, but the size of the two should be matched. A typical size is 8.5 millimeters in diameter.

GAS LASERS

Gas lasers were supposed to be the first built. In the theoretical paper written in 1958 by Arthur Schawlow and Charles Townes, they described an active material of potassium vapor enclosed in a tube. Maiman's ruby laser was created in mid-1960. Shortly after (2 or 3 months), Ali Javan and associates of the Bell Telephone Laboratories made the first gas laser. Gas lasers include three types: the neutral atom, the ion, and the molecular. Neutral atom lasers are electrically balanced and have an equal number of protons and electrons. An example of the neutral-atom laser is the helium-neon (HeNe) laser. This laser operates in the visible range of the electromagnetic frequency spectrum at a wavelength of 0.6328 microns. An ion gas laser has an active region whose atoms have missing electrons from their valence rings. An example of the ion gas laser is the singly ionized argon Ar+. This laser operates in the visible range of the electromagnetic frequency spectrum at wavelengths of 0.4880 and 0.5145 microns. A molecular laser has an active medium whose molecules have two or more atoms either of the same or different atomic species. An example of the molecular laser is the carbon dioxide laser. This laser operates in the infrared range of the electromagnetic frequency spectrum at a wavelength of 10.6 microns. Other prominent gas lasers are krypton, an ion laser (0.725 microns); helium-cadmium, an ion laser (0.4416 microns); and nitrogen, a molecular laser (0.3371 microns).

Gas Laser Operation   As a beginning, let us take a cursory look at the gas laser operation. Each gas laser has several things in common. A typical gas laser consists of an active gas medium within a glass tube, an excitation source, a feedback mechanism, and an output coupler. The active medium in the gas laser illustrated in figure 9–11 is helium-neon (HeNe). The gas is enclosed in a sealed glass tube called a *plasma tube*. The glass windows on each end of the plasma tube are made at an angle to transfer maximum light from the reflecting mirrors. The glass ends are called Brewster windows. One end of the plasma tube has a 100 percent reflective mirror while the other end is partially reflective and transmits most of the light energy back into the tube for more lasing action.

Now let's go through the operation and consider how it all comes together. Helium is added to neon (90 percent to 10 percent respectively) in a plasma tube for maximum pumping efficiency. Direct current is supplied to a negative and positive end (anode and cathode) of the plasma tube. Population inversion, stimulated emission, and lasing action take place. The beam of photons go through the Brewster windows on each side of the plasma tube. The beam strikes the 100 percent-reflecting mirror and is reflected back into the active medium. The beam strikes the partially reflecting (transmission) mirror and most of the beam is reflected back into the active medium. A small amount of the beam is released and coupled through an aperture for external use. If too much energy is coupled out of the transmitting mirror, the stimulated emission will decay and die.

The output of most gas lasers is extremely coherent, and continuous wave (CW) in operation. Most gas lasers are very low beam divergence and can easily be tuned. They also have low comparative prices and are of fairly rugged construction.

Figure 9–11 Gas Laser Operation (*Mike Anaya*)

A NEUTRAL-ATOM GAS LASER

Typical of a neutral-atom gas laser is the helium-neon (HeNe) laser. As previously discussed, the gas laser developed in late 1960 at Bell Laboratories was an HeNe laser. This first gas laser had a gas mixture of 90 percent helium and 10 percent neon. Neon gas, however, is the active medium in the HeNe laser. When helium is added to the neon, it aids in the pumping action. It seems that there is some difficulty in pumping neon atoms to their excited state. Helium atoms are very light in comparison to neon atoms and are easily excited. It so happens that the excited state of the helium atom is the same energy level as the neon atom's excited state. Therefore, when a helium atom in an excited state collides with a neon atom in the ground state it can transfer all its energy to the neon atom. These collisions happen frequently in the gas mixture at high pressure, therefore population inversion can take place and be maintained by pumping action.

Energy-Level Diagram for Helium-Neon Laser   The energy-level diagram shown in figure 9–12 is extremely simplified. It must be noted once again that the excited energy states of the helium and the neon atoms are almost the same. There are several of these excited states, but they have purposely been left off the energy-level diagram for better understanding. As the electrons are released from the dc power supply, they are drawn through the active region by the power supply's positive pole. The electrons collide with the gaseous helium atoms. The helium atoms are raised to an excited energy state. Those excited helium atoms in turn collide with neon atoms, and transfer all their energy to the neon atoms. The collisions occur more often at high pressure. This process is called *resonant excitation*. When this transfer happens often enough and at a fast enough rate to be larger than the decay rate to ground energy state, then population inversion can take place. In the figure, stimulated emission occurs between the excited energy state and an intermediate state. This stimulated emission is occurring at a wavelength of 0.6328 microns. Stimulated emission can occur in neon at several wavelengths. Others are 1.152 and 3.390 microns. Some lasers have these in the output simultaneously.

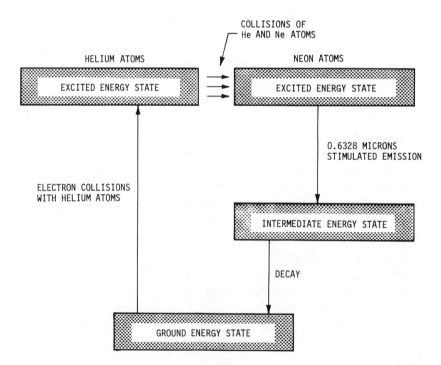

Figure 9–12   Energy Levels for Helium-Neon Laser (*Mike Anaya*)

The Plasma Tube   Plasma tubes are made in many different configurations. The line drawing in figure 9–13 depicts a simplified version of the HeNe plasma tube. A gas reservoir contains a mixture of helium and neon gases (HeNe). Just below the reservoir and connected to it is a capillary tube. The tube is where lasing takes place. The reservoir ensures that the supply of gas in the capillary tube does not fall low. For maximum power output, a constant pressure should be realized. The aluminum cathode tube is a hollow cylinder. On one end is a path for gas to the capillary tube. An element attached to the cathode lead prevents contaminants from entering the gas by collecting them electrically. The anode connected to the capillary tube is made of nickel. As in other lasers there are 100-percent-reflecting mirrors on one end of the tube and the partially reflecting output mirrors on its opposite end.

In operation a power supply supplies direct current to the anode (+) and cathode (−) for continuous-wave (CW) operation. The high-voltage pulse initiates the glow discharge (breaking down of the neon gas). Ionization of the gas involves removal of electrons from gas atoms. The ionization is synonomous to weak ionization. Stimulated emission can take place. The process of stimulated emission has been previously described in several passages of this section.

Figure 9–13   The Plasma Tube (*Mike Anaya*)

Helium-Neon (HeNe) Laser Applications   The HeNe laser envelope can be made of aluminum or glass. Aluminum of course is a stronger material. Problems in application are centered around the prevention of gas contamination. The HeNe laser is by far the most widely used. There are literally tens of thousands installed annually around the world. Use of HeNe lasers can be placed in three main categories. Low-power laser beams are routinely directed hundreds of feet away for alignment purposes. The construction industry accounts for the greatest numbers of all lasers in use today. It is used in trenches where it facilitates the setting of grades for sewer pipes. With a rotating beacon optical accessory the laser line can be rotated in a plane around a central point. It then provides a very convenient reference plane which can even be tied to the hydraulic controls of massive leveling equipment. With a simple low-cost optical accessory which fans the beam into a sector, lines can be projected against walls for leveling drop ceilings inside buildings, or for aligning lumber and planks in sawmills. On factory floors, the alignment of many parallel cylinders or rolls, like the ones found in production machinery in paper and steel rolling mills; the alignment of giant rotary kilns used in the cement industry; the alignment of complex assemblies such as turbine generators or linear accelerators; the alignment of another (invisible) laser beam (CO_2 or YAG laser) is finding widespread acceptance in machining materials such as metals, plastics, and others. In hospitals the HeNe laser is now used to precisely align patients for diagnosis or treatment by sophisticated x-ray equipment.

The HeNe laser can measure. Its light can be analyzed after it has been reflected off a variety of surfaces, after it has been shadowed by a number of objects, by the interference patterns due to its monochromaticity, or by the time it takes to travel over a given distance. Accurate dimensions can be recorded. Parts can be inspected for defects without part contact. This means no part damage, no gauge wear, no pressure contact error, and faster remote inspection. Lasers often perform better than their pneumatic and electronic counterparts in accuracy, resolution, and measurement range. They are not affected by radiofrequency interference, and their output is compatible with photodiodes, TV cameras, and other high-level output devices.

The HeNe laser efficiently detects surface defects (scratches, pits, breaks, burns, and missing threads on bolts). It can sort parts and measure thicknesses, diameters, profiles, flatness, and orientations. The HeNe laser also contributes to such analytical measurements as blood-cell sorting and small-particle measuring as required for pollution monitoring.

The major laser recording applications in use today are in facsimile transmission, telecopiers, computer line printers, oscillographic chart recorders, newspaper plate-making, industrial and supermarket label readers, optical character readers, and computer output microfilm. They "write" on a variety of media ranging from film to dry-silver paper, electrostatic film, and plain paper via electrostatic transfer techniques. Research is now developing new media with optimum response to the reliable and inexpensive red light of helium-neon lasers. The laser light offers significant advantages for recording: high speed and excellent resolution. More importantly, it can be directly controlled by a computer when coupled to a device such as an acousto-optic (A/O) modulator.

A Typical Neutral Atom Gas Laser   Typical of a neutral atom gas laser is the Spectra-Physics Model 155 (HeNe) laser shown in figure 9–14. This low-power (also low cost) laser has output power ranging between 0.4 and 0.76 milliwatts. The Model 155 laser operates at a 632.8 nanometer wavelength, with a beam diameter of 0.9 millimeter and a beam divergence of 1.0 milliradian. The power supply is contained within the laser head. A special neutral density filter within the supply assumes that the output will never exceed 1.0 milliwatt as established by the Bureau of Radiological Health (BRH) for Class II lasers. The Model 155 laser is used in laboratories and for educational purposes where low-power lasers are required. For user safety a key-actuated switch prevents unauthorized use. An emission-indicator light provides a visual signal to show that the laser is operating. A beam attenuator enables the user of the laser to block the beam at the output aperture without shutting down the laser.

Figure 9–14   The Spectra-Physics Model 155 Helium Neon (HeNe) Neutral Atom Gas Laser (*Courtesy Spectra-Physics, Inc.*)

AN ION GAS LASER

Ion gas lasers are usually operated with an active medium of argon or krypton. Argon operates at 0.5145 and 0.4880 micron wavelengths. Krypton operates at a wavelength of 0.7525 microns. Several years after the helium-neon (HeNe) laser was developed (probably 1963) the ion gas laser came into being. An American scientist, W. E. Bell, developed a laser using mercury ions. He found that it was fairly simple and easy to remove electrons from the outer shell of the mercury atom. Once the electrons were removed, the atom was left with a net positive charge. This atom then became a positive ion.

Ion gas lasers in their early stages were pumped by electron beams used also to ionize the gas. Heating of a metal cathode in turn boiled electrons from the cathode surface. The electrons were accelerated to high speeds while being focused into a dense, tiny beam. When the beam struck the gas atoms it knocked electrons from them and drove the new ions to an excited state where population inversion could take place. Problems with this system occurred because of poor concentration of atoms and high temperatures. The plasma tubes physically pumped gas from one end to another. A new tube fitted a return pipe which tended to equalize the end pressures. A magnetic field was used to concentrate the laser beam. These methods were short-lived, because the large current developed caused unbearable heat. Furthermore, the constant battering of positive ions on the walls of the tube caused physical damage. Units failed and new methods had to be developed.

Ion Gas Laser Operation   Modern ion gas lasers are extremely high power, easily controlled, and stable. The basic ion laser consists of a plasma tube, a resonator structure, and a power supply (see figure 9–15). The most important part of the ion laser is the plasma tube. It alone controls the output power and general performance of the laser. It is also expensive to replace and critical to the longevity of the laser. Modern ion lasers utilize graphite or beryllium oxide (BeO) for use in plasma-tube design. In the illustration you will note that the ion gas laser has a baffled plasma tube with a solenoid surrounding the tube. A reservoir is installed to the plasma tube and an ac power supply is tied to the anode and cathode. The ac power supply excites the anode and cathode, which in turn releases electrons. The electrons are accelerated through the gas in the plasma tube. The electrons bombard the gas atoms, which release electrons and become positively charged ions. The new ions raise to an excited energy state. A solenoid is placed around the plasma tube to concentrate the ion beam so that higher power can be developed. Baffles in the plasma tube allow for cooling. The concentrated beam produces extreme heat, while baffles act as a heat sink. Modern plasma tubes

utilize cooling water jackets. At the end of the plasma tube are Brewster windows
that allow maximum light transfer to optics. A prism external to the plasma tube
acts as a wavelength selector. The selector is tilted in respect to the plasma tube
axis. This causes high cavity losses except at one specific wavelength. This allows
the laser to be tuned to only one wavelength of those available from the gaseous
active medium. Highly reflective and partially reflective mirrors at either end of
the tube ensure the normal feedback operation. A gas reservoir attached to the
plasma tube maintains the tube with a constant supply of gas at a predetermined
pressure.

Figure 9–15   Ion Gas Laser Operation (*Mike Anaya*)

A GRAPHITE PLASMA TUBE

A typical graphite plasma tube is the Coherent Supergraphite™ plasma tube. In
figure 9–16, the plasma tube is installed within a Coherent S-series ion laser. You
will notice that the plasma tube is the bulk of the laser. This design produces
higher output power, which allows for lower operating current. Further, it eliminates drop-out noise from the beam. Drop-out noise is momentary but periodic
loss of the beam at the output. The graphite bore absorbs gas in a controlled
fashion and requires only periodic addition of gas. None of the electrical components are directly in contact with cooling water, therefore there is no electroly-
sis or possible shock hazards. The tube-water flow may come from most city water
supplies. The intense arc discharge of ion lasers causes erosion of material from

Figure 9-16 A Graphite Plasma Tube within an Ion Laser (*Courtesy Coherent, Inc.*)

one section of the bore and its deposition in another section. The deposited material will eventually cause constriction and decrease in laser output. The bore of the graphite plasma tube is segmented to compensate for changes in bore diameter. That is, the bore is enlarged in areas of heavy buildup. Brewster windows are made of crystalline quartz to prevent the formation of color centers. Color centers may contribute to degradation of output power. Discharge tubes have extensive heat shielding internal to the vacuum envelope around the anode and cathode. These shields direct the heat into the water jacket away from the resonator.

In figure 9-17A, a unique tapered cathode is shown. The traditional cathode is a tungsten spiral with a uniform cross section. In the traditional cathode, discharge is concentrated at the end of the cathode nearest the bore and the unequal distribution of current causes excessive heating of the front ring of the cathode. In the tapered design, the end nearest the bore has a much larger cross sectional area than the end away from the bore. Current is therefore more evenly distributed and cathode temperature is more uniform. This design prevents cathode fatigue and extends its life.

An ion laser operates with an intense arc discharge which accelerates ions into the walls of the plasma tube where they are buried and cause a drop in gas pressure. This gas must be replaced or performance will deteriorate. In figure 9-17B, a Coherent dual-valve refill system illustrates refill of the system by metering the gas into the tube in precise increments. It thus allows optimum pressure to be maintained. Automatic pressure controls allow the pressure to be monitored and gas to be added as required. Monitoring is accomplished by a direct-reading pressure meter.

Ion Laser Bore Contouring   Because of the cost of plasma tubes, the tube is generally around one-third the system price. It stands that if the tube is that expensive then ion laser manufacturers are in constant research to increase the reliability of the tube. Figure 9-18 illustrates some configurations of plasma-tube bores within ion lasers. In the figure 9-18A, a traditional bore is shown after only

(a) A tapered cathode

(b) Dual value refill system

Figure 9–17 Cross Sections of the Coherent Graphite Plasma Tube (*Courtesy Coherent, Inc.*)

a few hours of use. In figure 9–18B, the same bore is shown after 1000 hours of use. It is rather obvious that in the latter the bore material has sputtered away in some areas and deposited in others. This process will occur in all ion lasers regardless of bore material. The constrictions behind the cathode and the anode taper sections squeeze the laser mode, causing diffraction and loss of output power. This can be identified in only a few hundred hours of operation.

The graphite-bore design relieves this problem. In figure 9–18C, a contoured bore is shown after a few hours of operation. In figure 9–18D, a contoured bore is illustrated after 1000 hours of operation. The areas where buildup is expected is contoured and enlarged in manufacture. As sputtering occurs, these areas are slowly filled up but do not occlude the beam. The contoured bore may extend the life of the plasma tube to as much as 6000 hours.

(a) Traditional Bore (few hours operation)

CONSTRICTION   ENLARGEMENT   CONSTRICTION

(b) Traditional Bore (1000 hours operation)

(c) Contoured Bore (few hours operation)

(d) Contoured Bore (1000 hours operation)

Figure 9–18   Bore Contouring (*Courtesy Coherent, Inc.*)

Beryllium Oxide (BeO) Plasma Tube   The BeO plasma tube is used because of its properties of high thermal conductivity, its excellent strength at elevated temperatures and its low gas absorption rate. Thermal properties allow immediate shutdown of the laser and the cooling water. Heat stored in the plasma tube at shutdown is not sufficient to raise the cooling water in the tube to boiling temperature. The tube is conduction-cooled. This means that all the heat is carried away by the cooling water. Heat does not effect resonator alignment, thereby ensuring high, stable output power. There is no need to make continual optic adjustment to compensate for heat.

The low gas-absorption rate of this plasma tube means less-frequent gas fills and more consistent output power for longer periods of time. Use of beryllium oxide (BeO) material as a bore material minimizes pressure changes in ion lasers.

The Coherent COMPOSITON® Resonator The resonator in figure 9–19 comprises the laser mirrors and the support structure that maintains mirror alignment. The resonator structure is responsible fo the amplitude and frequency stability of an ion laser system. For maximum performance the mirrors must be aligned to within a few microradians. A small deviation of as small as 10 microradians can cause changes in output power of as much as 20 to 30 percent. The Coherent model uses three orthogonally mounted solid-fused quartz rods as the main support element. The low expansion coefficient of the rods results in minimization of angular misalignment caused by differential heat loading and of frequency drift resulting from cavity expansion. Each quartz rod is surrounded by an aluminum heat shield with a layer of air for insulation. The shields stabilize the rods against temperature fluctuations and maintain even temperature across and along the rods. Shields are thermally interconnected by cross members and heat-flow plates to ensure that all rods have a uniform thermal environment.

Figure 9–19 The Coherent COMPOSITON ® Resonator (*Courtesy Coherent, Inc.*)

To prevent the disturbing of mirror or plasma-tube alignment, the support structure is kinematically mounted by a combination of spherical bearings and leaf springs so that thermal and mechanical stresses on the laser housing are decoupled from the resonator.

The Spectra-Physics STABILITE™ Resonator Structure The two most important considerations in the design of a laser resonator are the angular and length stability of the optical cavity. These determine the ultimate amplitude and frequency stability of the laser. The best technique for achieving optimal performance is to select a material that has a very low thermal coefficient of expansion and to thermally isolate this material from ambient temperature changes and heat-generating components in the laser. Another technique is to select a material that adjusts uniformly to temperature changes so that optical alignment is always precisely maintained. The Spectra-Physics resonator in figure 9–20 uses both techniques in its design. The aluminum extrusion structure utilizes the property of excellent thermal conductivity to produce amplitude stability. The structure is kinematically mounted so that thermal and mechanical stresses cannot be coupled into the resonator to disturb the mirror or the plasma-tube alignment. Three spherical bearings kinematically isolate the entire optical cavity from mechanical stress applied by the outer case, end plates, and the feet of the laser. These bearings also relieve thermally generated mechanical stress that might originate in the resonator structure.

Figure 9–20 The Spectra-Physics STABILITE™ Resonator Structure (*Courtesy Spectra-Physics, Inc.*)

Ion Laser Applications The ion laser has a multitude of applications which are dependent on the power required. Ion lasers are used in the fields of holography, spectroscopy, data recording and retrieval, eye surgery, dye laser pumping, biological and cell-sorting experiments, and printing plate exposure systems. High-power ion lasers are ideally utilized for isotope separation. The ion laser is again one of the most useful tools for general laboratory work that requires a strong, versatile light source.

A Typical Ion Gas Laser Typical of the gas ion lasers is the Coherent S–Series CR–2 seen in figure 9–21. This laser is an argon ion laser. The laser operates at visible blue, green, and ultraviolet wavelengths. It uses a graphite plasma tube as illustrated in figure 9–17 and a resonator as shown in figure 9–19.

Figure 9–21 The Coherent S Series CR-2 Gas Ion Laser (*Courtesy Coherent, Inc.*)

The laser is powered by 208-volt alternating current 3-phase input power. The configuration shown has a built in plasma tube, resonator, and power supply. The power supply must be current-regulated. Optics are also within the case shown. A Brewster prism allows selection of individual wavelengths. Bimetallic compensation of the support structure ensures stable operation in a changing-temperature environment. Output beam light is sampled with a silicon detector and a feedback loop for regulation. The support structure mount is designed to compensate for power fluctuations due to angle changes of the unit. A separate power meter is used to monitor power output.

A MOLECULAR GAS LASER

The molecular gas laser was developed in 1965 by C. K. N. Patel, a scientist at Bell Telephone Laboratories. The molecular laser is the most powerful of the gas lasers and most efficient. As early as 1967, North American Aviation had produced a CO_2 laser with an output of 4000 watts. The molecular CO_2 gas laser is so powerful that it can be used for drilling holes in metal, welding, and even rock crushing. The molecular nitrogen (N_2) gas laser is used to pump tunable dye lasers. The molecular gas laser utilizes the molecule for basic operation. You may recall from high school physics that a molecule is made up of atoms which may or may not be the same. An example of a molecule you may be familiar with is water (H_2O).

There are two primary molecular gas lasers. These are carbon dioxide (CO_2) and nitrogen (N_2). The CO_2 laser operates around 10.6 microns in the infrared range of the spectrum. The nitrogen laser operates at 0.3371 microns wavelength in the near ultraviolet range. CO_2 lasers can be continuous wave (CW) with peak power to 1000 watts, repetitively pulsed at with peak power to 3000 watts and with Q-switching may reach peak power of 100,000 watts. Nitrogen molecule gas lasers (also called *superradiant*) operate a peak power of 100,000 watts.

Molecular Carbon Dioxide (CO_2) Gas Laser Operation   Molecular gas lasers operate in much the same manner as the neutral-atom gas lasers. Molecular gas laser systems are excited by electrical current flow in the direction of the light source, the same as in the HeNe laser. In the CO_2 laser, a mixture of carbon dioxide (CO_2) and nitrogen (N_2) is used. The nitrogen serves as the exciter, similar to the use of helium in a neon laser. Nitrogen molecules are easily pumped to an excited energy state by electrical discharge. In turn, carbon dioxide molecules are raised to an excited energy state by collision with nitrogen molecules.

The original molecular gas lasers had the active gas such as CO_2 flowing between reflection mirror ends. The flow would encounter nitrogen molecules that were previously excited by some form of electrical energy. The two gases would clash and energy stored in the nitrogen atoms would transfer to the CO_2 molecules. Lasing action could take place. The flowing gas CO_2 laser develops greater power with higher rates of gas flow. Carbon dioxide (CO_2) was later found to be able to function standing still and the gas flow was no longer required.

Another carbon dioxide laser is the TEA CO_2 laser. The TEA CO_2 laser is excited at several points perpendicular to the optical flow between reflective mirrors. TEA is an acronym for *t*ransversely *e*xcited at *a*tmospheric pressure. The TEA CO_2 laser has a high-voltage direct-current power supply that charges a large capacitor. The capacitor is discharged into the gas in the plasma tube at predetermined times. Switching action controls the capacitor charge and discharge times and therefore the laser pulse.

Molecular Nitrogen (N_2) Gas Laser Operation   The molecular nitrogen (N_2) operates in much the same manner as the TEA CO_2 laser. Nitrogen molecules are brought to an excited state by a high-speed electrical discharge from capacitors. The discharge channel can be so fast that mirrors are not required for reflection. In most cases the mirrors are used and a superradiant output results. The N_2 laser does have limitations in that its pulse duration is short at 10 nanoseconds. Furthermore, nitrogen gas has a low energy-storage capacity. Capacitors may be charged in series and discharged in parallel or charged in parallel and discharged in series. The latter, called a Blumlein circuit, provides higher peak currents and pulses of 3 to 5 nanoseconds duration. This is compared to the 10 nanosecond pulse of the parallel discharge method.

N_2 lasers use either channelized or open discharge channels. In the channelized design, a double-discharge beam is projected along the outer walls of the discharge channel. In the open-discharge channel, the channel is made large and the beam is projected through the center of the channel. The double-discharge beam is not suited for dye laser pumping because of limited output energy.

As a dye laser pumping source the nitrogen laser has no equal. The most important parameter that governs dye laser efficiency is the pulse duration of the nitrogen laser. For all cavity lengths and pumping peak powers, dye laser efficiency increases with pumping pulse duration. A long pulse is always more efficient than a short pulse.

Molecular Gas Laser Applications   Because of the high power of the molecular gas laser it has become an important tool in machining operations. CO_2 lasers are used for a great number of cutting, drilling, and welding tasks in materials which include metal, rubber, and plastics. Beams of the CO_2 laser can be focused to as small as 0.005 inch for drilling and cutting. Most materials are instantly vaporized and may be immediately removed. There is no debris from laser drilling and cutting. There is no tool contact, meaning no tool wear or replacement. Metal welding includes materials of such hardness as monel metal and titanium. Nitrogen lasers are primarily used as pumping devices for dye lasers. More on these applications will follow later in this chapter.

A Typical Molecular Gas Laser   The Coherent Model 42 in figure 9–22 is a molecular gas CO_2 laser system. It was designed for laboratory and industrial uses. Its output is at 10.6 micron wavelength with power up to 50 watts, which makes it ideal for general materials research. The infrared beam can be focused to as small as 0.12 millimeter in diameter for precise welding, cutting, and drilling of a wide variety of metals. Materials are instantly vaporized and easily removed from the material. The system can be operated from 115- or 230-volt alternating current single-phase power at 60 hertz. Pulse rate of the system is 1 kilohertz. Pulses are 100 microseconds wide. The Coherent Model 42 has several distinguishing features. These are: A catalytic converter reduces gas consumption by recombining dissociated molecules. A rotating lens moves the laser beam in a circular pattern for precision cutting of large holes. A special coaxial gas jet bathes the processing area to either neutralize or oxidize a specific environment. The Coherent Model 42 comprises a discharge tube, resonator, power supply, and

Figure 9–22   The Coherent Model 42 CO_2 Molecular Gas Laser (*Courtesy Coherent, Inc.*)

gas-flow system. The discharge tube and resonator are mounted in the laser head, a package separated from the power supply and the gas-flow system.

LIQUID LASERS

Liquid lasers, also called dye lasers, utilize organic dye as an active medium. The advantage of using a dye laser over others is that the dye laser can be tuned over a wide variety of wavelengths for each dye used. The variety of the dyes literally extends over the entire visible range of the spectrum with some dyes in the near-infrared ranges. This latter attribute makes the dye laser a highly respected tool in biomedical research.

Some of the dyes utilized in liquid laser operation and their lasing range are as follows:

Dye	Range
Carbostyril	419 to 485 nanometers
Coumarin	435 to 565 nanometers
Sodium fluorescein	538 to 573 nanometers
Rhodamine	540 to 690 nanometers
Cresyle violet/rhodamine	675 to 708 nanometers
Nile blue	710 to 790 nanometers
Oxazine	695 to 801 nanometers

The center of the range provides the user maximum dye absorption. Optimum dye laser performance is achieved by spectrally matching the dye laser pump with the absorption band (range) of the dye, then tuning the concentration of the dye for maximum absorption. The dye is generally mixed in an alcohol or water solution or some special solution which provides a better mix. Liquid detergent is often added to aid in lasing performance.

Dye lasers can be continuous-wave (CW) or pulsed. In operation dye lasers are pumped by gas lasers or flashlamps. A few are pumped by ruby or Nd:YAG solid lasers.

Liquid Laser Operation An optical schematic of the Coherent Inc. Model 590 Dye Laser is shown in figure 9–23. The Model 590 laser is continuous-wave (CW). The CW dye laser operates by rapidly moving dye molecules through a CW laser pump beam, spatially coincident with the optical cavity of the dye laser. Each molecule in effect sees a pulse of light as it passes through the pump beam.

A dye is selected that will provide the spectral range desired. CW dye lasers span the spectral range from 420 to 800 nanometers. The dye is dissolved

Figure 9–23   Dye Laser Operation (*Courtesy Coherent, Inc.*)

in ethylene glycol. The dye is squirted in a flat stream into the laser cavity at the Brewster angle. The dye is circulated with a gear pump through the cavity, a filter and a reservoir at a velocity of around 6 or 7 meters per second.

The dye solution is excited by a beam from an argon ion laser. The beam is focused to a specific diameter by a focal length mirror M4. The input mirror M4 directs the beam to the cavity and the dye stream. The optical cavity consists of the three mirrors M4, M2, and M1. Mirrors M2 and M1 are highly reflective. M2 is the end mirror which serves as reflector and a collimator to the output coupler mirror M3. Mirror M1 is a reflector. The output coupler mirror M3 is a flat mirror with a typical transmission of 4 percent.

Tuning the CW dye laser wavelength is accomplished with the birefringent filter placed in the laser cavity. The filter consists of three quartz plates of different thicknesses aligned in the cavity at Brewster's angle and rotated to change the wavelength of the transmission.

Typical Tuning Data   The Coherent Model 590 laser may operate with fourteen different dyes at wavelengths that range from 420 to 800 nanometers. (See figure 9–24.)

A Flashlamp-Pumped Dye Laser   Not all dye lasers are pumped by another laser. Typical of these nontraditional lasers is the Phase-R Company DL

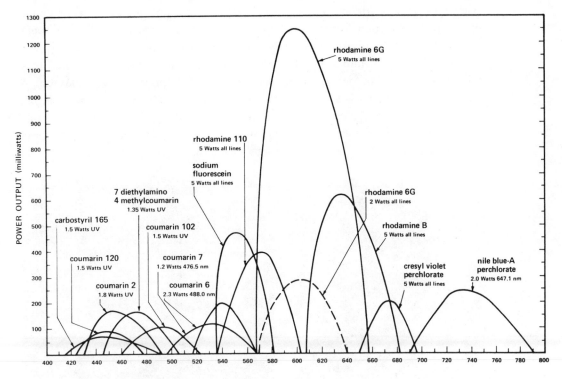

Figure 9–24 Typical Dye Laser Tuning Data (*Courtesy Coherent, Inc.*)

series lasers. These dye lasers are pumped by a coaxial flashlamp. They are, as one would expect, less expensive than other dye lasers in that flashlamps are cheaper than lasers as pumping sources. Its basic function is thus: when the lamp is discharged, the dye solvent liquid that flows continuously through the center of the lamp is excited uniformly from all directions. This ensures the best optical coupling efficiency, the best overall electrical efficiency and the lowest threshold that can be obtained from any flashlamp pump method. With minor modifications the Phase-R DL series lasers will pump dyes into the ultraviolet range as low as 3400 angstroms and into the infrared range up to 9600 angstroms.

A Typical Liquid (Dye) Laser A typical continuous-wave (CW) laser is the Coherent CR–590 seen in figure 9–25. The basic system consists of a resonator structure, a dye-flow system, and a tuning element. The configuration utilizes a three-mirror laser cavity (see figure 9–23). The laser is pumped by an argon or krypton gas ion laser focused into the dye stream. The dye stream is inserted at

Brewster's angle in the cavity. All the elements of the optical cavity are mounted on a 2-inch Invar bar. The rigidity of the bar ensures alignment stability. A stainless steel jet nozzle injects an optically flat stream of dye into the laser cavity. The jet is hard-mounted into a swage fitting. The jet assembly is then mounted to a micropositioner. The jet is completely enclosed to prevent dye splashes. A neutral-density filter allows a view of the jet. A crystalline quartz birefringent filter provides smooth, precise tuning. When pumped by the argon laser pump, the output is around 690 nanometers. When pumped with a large krypton laser the output range can be between 390 nanometers to above 1.0 micron.

Figure 9–25 The Coherent CR-590 Dye Laser (*Courtesy Coherent, Inc.*)

A Built-In Laser Pump Dye Laser Most of the Sun Technology Chromaser series lasers have a built-in helium-neon laser system that is adjustable to be coaxial with the pulsed laser beam. The helium-neon beam can be used for internal alignment of the pulsed tunable laser system or it can be used externally as a target designator for the pulsed tunable laser. Bandwidth coverage can be extended from 3800 to 7800 angstroms by changing reflectors on the system. A mono-chromater is installed on the laser system to verify the operating wavelength. The Chromaser series have unique applications. They are used along with micro-scopes to support the biological sciences.

The microscopes were chosen to have optics that were compatible with highly energetic laser beams and have the quality of not affecting the wave front of the passing beam. This allows the microbeam to be always held in focus. As the operator varies the specimen around the focal point of the objective they can easily selectively destroy microorganisms or bacteria in the horizontal or vertical plane. The operator is also able to examine a specimen and fire the laser while maintaining constant vigil of the specimen during firing. This is especially im-portant to a biologist when live specimens are being irradiated. Most organisms, particularly bacteria, are transparent, therefore absorption is very low. In order to destroy the bacteria, cells, or organisms, the laser must have high energy and peak power. The pulsewidth of the Chromaser is less than one-third of a microsecond, which provides billions of watts of pulse power at the objective focal point. During tests, 0.3-micron holes have been drilled in 1.0-micron bacteria. Other tests have selectively destroyed or vaporized 1.0-micron bacteria. Single cells in worms were selectively destroyed with the worm remaining alive.

OPTIC SETS

Each laser has a set of optics that is compatible to that laser. Depending on the laser, this set of optics may include optics such as reflectors, windows, beamsplit-ters, polarizers, etc.

Mirror radius has an effect on output power. Decreasing the mirror radius will decrease the fundamental mode (TEM_{00}) power while increasing multimode power. Different mirror material may cause color centers on windows. Tempera-ture change or stress-caused angular fluctuations of mirrors which in turn causes a change in output power. All these things and others make it imperative to select the designed optic set.

Figure 9–26 depicts a typical lens set. This set is for a carbon dioxide laser. The optics here are manufactured from high-quality germanium. The ideal optical material has low absorption, low strain, and high thermal conductivity. In addition, the material should be hard enough to take a good optical polish and should support high-durability dielectric coatings. Finally, the material should not go into thermal runaway. Thermal runaway is the increase in optic absorption as temperature increases.

Figure 9–26 Optic Sets (*Courtesy Coherent, Inc.*)

Choice of optic sets is a matter of what is demanded by design. In general these design demands fall into several categories:

1. Surface quality
2. Area of exposure
3. Surface flatness/accuracy
4. Radius of curvature tolerance
5. Specifications of polarization

Substrates The material that optics are made from is called a *substrate*. A substrate is simply the base material prior to modifying it to do a particular job. Substrate material is selected for specific properties. These properties are absorption, thermal conductivity, and water solubility. After the material has been selected for the substrate, general specifications come into play. Some of the general specifications are surface conformity, surface quality, edge finish, and diameter and thickness tolerances. Surface quality may also be defined as the wavelength realm in which the optics function efficiently.

Substrates in the uncoated state generally come in two categories. These are infrared and ultraviolet to near-infrared. Infrared substrates are made of germanium, gallium arsenide, cadmium telluride, zinc selenide, and sodium chloride. Germanium has become the prime material used in infrared substrates. It is relatively inexpensive, takes a very good polish, and is available in large sizes. With adequate cooling germanium operates well. With higher power, heat becomes a problem. High-power applications demand that gallium arsenide or zinc sellenide be used.

Gallium arsenide allows the infrared laser system designer to work with higher power densities than previously possible with germanium. The material does not become absorbing or go into thermal runaway until quite elevated temperatures are reached.

Zinc selenide grown using vapor-phase deposition techniques has recently become available in sizes sufficiently large and at prices reasonably low such that it merits consideration for use as CO_2 laser optical elements.

The ideal optical material would have low absorption at high thermal conductivity, and low strain. In addition, the material should be hard enough to take a good optical polish and should support high-durability dielectric coatings. Finally, the material should not go into thermal runaway; that is, it should not show increased absorption, as the temperature of the optical element rises. Zinc selenide approaches these ideals. Salts, including sodium chloride and others show very low absorption figures. Unfortunately, salts are affected by moisture, do not take a good polish, and exhibit quite low thermal conductivities.

Semiconductor materials such as germanium, gallium arsenide, and cadmium telluride have been used with success as optical elements for CO_2 lasers. These materials possess good thermal conductance characteristics and have reasonably low absorption figures. All of the semiconductors are, however, to a greater or lesser degree susceptible to thermal runaway.

Uncoated substrates used in the ultraviolet to near-infrared range are made from optical-grade fused silica or some special variation. They tend to be thick to ensure stability. As in infrared substrates, they must be able to maintain excellent transmission qualities while providing very low temperature distortion.

In figure 9–27, a typical laser mirror substrate is illustrated. Note that measurements required by specification include the diameter and thickness. R_1 and R_2 represent the radii of the first and second mirror surfaces. S_1 and S_2 are sides one and two respectively. The wedge angle (S_1 and S_2 parallelism) is also a required specification. Details on substrates types are provided in chapter 6.

Coating of Optics Optics are often coated with dielectric materials to optimize their performance. With the use of these coatings there evolves a wide variety of coated optics capable of selecting a large range of wavelengths for operation. The dielectric materials are generally more sturdy than the substrate itself;

Figure 9–27 Substrate Specifications (*Courtesy Coherent, Inc.*)

that is, the surface is harder and it can withstand extremes of temperature and humidity. Substrates coated with dielectrics can be immersed in water. The coated surface is cleaned with acetone or methyl alcohol.

Coatings for dye laser mirrors are specifically selected to match the designed performance of the laser. Figure 9–28 illustrates a coated optic under extreme magnification. The photo shows the multilayer dielectric coating after scribing it with a diamond. The scale is 17.9 millimeters = 1 micron.

Figure 9–28 Multicoatings of Dielectric Material on an Optical Substrate (*Courtesy Spectra-Physics, Inc.*)

TRANSVERSE ELECTROMAGNETIC MODES (TEM)
LASER BEAM PROFILES

There are dozens of beam profiles with which lasers operate. Several of these patterns are illustrated in figure 9–29. These patterns are perpendicular to the beam in its direction of operation. Longitudinal modes are in the direction of wave propagation in a spatial Z-axis orientation. The field which represents the light beam consists of all three spatial coordinates X, Y, and Z. The Z-axis is perpendicular to the X–Y coordinates, which makes the X–Y fields perpendicular to the direction of laser beam travel.

The basic mode of lasers is called the *fundamental mode* or the *transverse mode*. In figure 9–29a, this mode is illustrated. The pattern is labeled TEM_{00}. It would seem that all laser beam profiles should take this pattern and rightly so. Once the bore characteristics of a laser have been decided upon, the beam profile is determined by spacing and curvature of reflective resonator mirrors. The mirrors are chosen to ensure a TEM_{00} pattern at the longest wavelength. The mode diameter varies with the square root of the wavelength. Variations in the cavity mirror diameters can cause wavelength lines to be TEM_{00} or fundamental mode.

Subscripts represent the number of times the magnetic field crosses the X, Y, and Z axes of the laser beam. They also represent the number of modes in the standing-wave pattern along each axis. (See figure 9–6.)

Figure 9–29b is called a *doughnut mode*. It is the superposition of one mode on another, with the small circle as the second mode. This mode, TEM_{01} can also be caused by contamination present on a mirror face. Figure 9–29c, d, and e

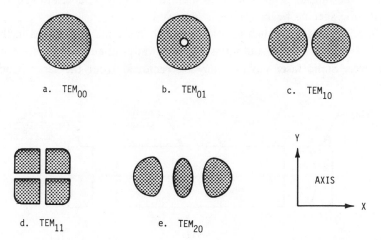

a. TEM_{00} b. TEM_{01} c. TEM_{10}

d. TEM_{11} e. TEM_{20}

Y

AXIS

X

Figure 9–29 Transverse Elecromagnetic Modes (TEM) Laser Beam Profiles (*Mike Anaya*) are representative of multimodes. A cavity that operates in multimode operation is said to be "lossy," because modes other than the fundamental mode TEM_{00} do not have maximum power density. The reason for multimodes is the inability of the laser to focus due to phase changes across the beam and dispersion or spreading out of beam power. The figure simply illustrates several of the beam profiles. In actual practice these beam profiles may take up a multitude of patterns.

LASER OPTICS ALIGNMENT

With the understanding that alignment is a fundamental requirement to all lasers we can make two general statements. First, to establish maximum power output the laser mirrors must be perfectly aligned. Second, alignment is said to be at its best when an imaginary line can be drawn perpendicular to the center of each mirror. When these lines are in coincidence with each other and with the axis of the active region the cavity is said to be aligned. (See figure 9–30.) This fundamental alignment is dependent upon the laser in need of alignment. Some essential requirements fit all lasers. The cavity of the laser along with its optics must be placed on an optical bench that has adjustable supports. In the case of some lasers, alignment can take place in the laser mount because adjustment devices are built in. Power is applied to the cavity and adjustment screws are used to mechanically move the laser optics in the horizontal, vertical, and directional (X, Y, and Z) axes. Again this is dependent on the laser under alignment. An index card may be used for visual output lasers to align for a sight profile of the laser. The card is placed in the output beam line while adjustment takes place. Adjustment continues until the specific beam profile is achieved.

  Special equipment such as a laser power meter may be placed in the laser beam path to measure power output as alignment adjustments are made to monitor output power. Some laser cavities may be removed from the laser head. This

Figure 9–30 Laser Optics Alignment (*Mike Anaya*)

cavity (such as in the helium neon (HeNe) laser) may be aligned with another laser. The HeNe plasma tube is mounted on an optical bench on supports.

An operational laser is mounted on a three-way adjustable support on the optical bench so as to direct a beam through the center of the plasma tube. As the beam is projected through the cavity, a black index card is placed on the opposite end of the plasma tube under alignment. The card will show the beam profile if it is partially aligned. Alignment of the operational laser with the plasma tube is necessary. If the index card shows a profile, the operational laser beam is not hitting the walls of the plasma tube. The 100-percent-reflectance mirror of the laser is placed in a mount on the optical bench. This mirror is adjusted in the same manner as the operational laser with mount adjustment screws. An index card is used to show adjustment.

DYE LASER TUNING

The dye laser has the capability of adjusting its output beam to a variety of wave-lengths, depending on the type of dye being used. Just as there are many laser types, there are also many tuning methods. One of the best methods for tuning is by use of the birefringent filter. Typical of the use of the birefringent filter is with the Coherent Model 590 dye laser. The birefringent filter is insensitive to cavity misalignment and vibration. Prisms or etalon tuners are sensitive to vibration.

The birefringent crystalline quartz filter is placed in the dye laser at Brewster's angle. The filter is made of three crystalline quartz plates with a thick-ness ratio of 1:4:16. The thinnest plate determines the spectral range, while the thickest plate determines the bandwidth. The plates are oriented so that the optic axis of each crystal is in the plane of the face. The axes of all the crystals are in parallel with each other. The wavelength changes because the index of refrac-tion changes. In this manner the laser output wavelength is tuned.

The laser is different from other lasers in that it is the filter that deter-mines the wavelength and not the alignment of the filter in respect to the resonator structure. The filter can be replaced without realigning the laser. There is no reflective coating, therefore insertion loss is minimal. Finally, one tuning element is suitable for all wavelengths.

Figure 9–31, shows the effects of using three plates with a thickness ratio of 1:2:4. The plates are positioned at Brewster's angle inside the dye laser. Wave-length is achieved by rotating the plates around an axis normal to the plates. Theta (θ) represents the angle between the crystal axis and the polarization direc-tion. The curves represent the wavelength at 45° and 60°.

Tuning is usually precisely controlled by a calibrated micrometer drive.

It is possible to set the filter to within 1.0 angstrom of a particular wavelength. Other methods of tuning utilize a tuning wedge, an etalon, or both.

Adjustments are made on the high-reflectance mirror mount until the incidence beam and the reflected beam are superimposed. The partially reflecting output mirror is placed in its mount on the opposite side of the plasma tube and adjustment is made in the same manner as with the 100-percent-reflectance mirror. After alignment, the alignment laser is removed and an optical power meter is placed in front of the laser output mirror. All adjustment screws are adjusted to establish peak power output.

Again, alignment procedures generally follow similar patterns, but each laser has its own special peculiarities. Since output power and alignment are indeed essential, it behooves the user to adhere to alignment procedures developed by the manufacturer.

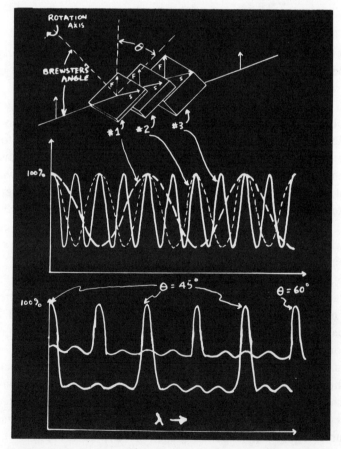

Figure 9–31   The Birefringent Filter (*Courtesy Coherent, Inc.*)

SPECIAL LASER APPLICATIONS

Holography   Holography was invented by Dennis Gabor, a professor at London University in 1947. Holography is the production of three-dimensional displays with the use of laser beams. Three-dimensional pictures have long been the desire of the motion picture industry, photographic technicians, inventors, and entrepreneurs of that art form. In the early 1950s, the 3D motion picture subjected the public to the wearing of special glasses in order to view pictures in the third dimension. It did not take hold. Even prior to that a special set of glasses was used to look at a double picture which gave a three-dimensional appearance. These were not acceptable after a time. So the search went on to find a 3D device that would allow you to look behind an image without the special help of spectacles or other devices and with no effort on the viewer's part except for looking. We shall briefly describe holography and then discuss its applications.

When light shines on an object the object either absorbs the light or reflects it. The reflection of the light from the object takes the shape of the object in three dimensions. A single eye sees only the intensity and the color of this light, as does a photographic film. Distance does not enter into the picture that one eye or a photograph sees. Two eyes, however, view the object at two angles. The human brain provides the intelligence to compute the various terms of distance. We see 3D because of our brain.

It is possible, with holography, to use a third component of light, its phase, to produce a three-dimensional record of incident light. The process involves two basic steps. These are the development of a hologram of an image and the reconstruction of the image. A hologram is made by reflecting half of a laser beam as a reference with a plane mirror on to a photographic plate. The second half of the laser beam is reflected off the image onto a photographic plate. The photographic plate records the interference pattern (phase relationship) of the two half-beams on the plate. The result is a photograph. This photograph contains intensity and phase variations. The record is called a *hologram* and is unlike any photograph you have ever seen. Several holograms can be recorded on one plate if the light beams are reflected at different angles. The hologram in itself is a complex-looking mass of patterns that represent the image that was reflected. To retrieve the image, a laser beam is driven through the hologram in the same direction as the original reference beam. If you look at the hologram from its opposite side a three-dimensional form will be obvious. If you move your head, you can observe the image from another angle just as you would view any three dimensional real image.

As you may have concluded, holography, though a fairly simple concept, requires a great deal of technical expertise. The development of an image to a hologram and then the reconstruction of the image is a new technical field and is worthy of a life's career. Moving objects are being recorded and displayed by holography. Holographic television was patented as early as 1940 by Professor Fritz Fischer. In industrial manufacturing, holography is used for measuring vibrations and small deformations in metal. In early 1980s, making a hologram will be just like shooting a roll of film, at least so say the experts.

Welding, Drilling, and Cutting   Lasers are not common in every small plant in the country to do the jobs of welding, drilling, and cutting because of the initial expense. The time is coming when the precise nature of the laser beam coupled with its intense power will be more than an expensive dream. The laser can confine high intensity. This and the characteristic of noncontact is extremely important where delicate parts have to be joined; contamination must be eliminated and thermal distortion cannot be tolerated.

Industrial welding has utilized two laser types, the CO_2 gas laser and the Nd:YAG solid-state laser. The CO_2 laser is used where high power and high efficiency are required. The Nd:YAG laser is used where high precision and minimum thermal damage are demanded. The Nd:YAG is also utilized where difficult welds are needed. Difficult welds are those which involve high-melting points, high conductivities, or metal mismatches. There are two basic welding modes of operation utilizing the laser. These are conduction and penetration. In conduction welding, the laser beam is impinged on the material at a power density that allows fusion but does not cause vaporization. Energy of the beam is conducted to the base metal. The fusion zone provides an aspect ratio of unity (1:1 ratio of depth to width). Penetration welding takes place with laser power necessary to drill a hole in the material and the hole is propagated through the material to form a fusion zone. Vaporization does take place. Aspect ratios are larger because the energy is not limited to the surface weld as in conduction welding.

The pulsed laser can take the place of hand soldering. Soldering is slow, tedious, and messy. Furthermore repeatability is almost impossible to achieve and quality control is difficult. The pulsed laser can allow welding in close proximity to delicate components. Seam welding and spot welding are both used without changing laser energy levels. Production lines in laser welding now weld such things as pacemakers, instrumentation, gyroscope bearings, and sensor probes.

Laser drilling can offer tremendous savings for the manufacturer because of its speed, but there are limitations to its use. Precision holes are not possible in metals more than a few thousandths of an inch thick. The diameter of the hole is uncontrollable and the face of the hole tends to become funnel-shaped. Thin layers of recast metal cling to the side of the holes. This recast metal is structurally different than the original metal. It may be harder and susceptible to cracking. Despite the problems, lasers are used to efficiently drill holes. Diamonds are drilled in minutes with a laser, where hand drilling required hours. Mass production and speed are the attributes of the laser drill. Holes that need a better finish are sized or polished after the laser operation.

Cutting operations have the same general characteristics as drilling operations. Cutting, however, may be the wrong term, for the laser does that job by heat rather than physical force. Temperatures for cutting may be up to 150,000°C. Vaporization does take place. Lasers have been used in the clothing industry to cut material.

Wood and Ceramics   The CO_2 laser is used in cutting, drilling, and etching of ceramics and wood. Energy from the laser is focused as a beam through a lens onto the ceramic or wood. This energy is high intensity in the megawatts-per-centimeter range. The ceramic and wood absorbs the high energy in localized areas. The material quickly melts and vaporizes.

Resistor Trimming   The Nd:YAG laser is utilized for resistor trimming along with several others such as the xenon ion laser and the CO_2 laser. Resistor trimming is categorized as thick and thin film technologies. The Nd:YAG laser is used throughout thick-film trimming because of its high repetition rate and high pulse power. The xenon ion laser is utilized with thin-film trimming because it is highly accurate and can be used in small areas. Resistor trimming with a laser is accomplished by focusing the laser beam on the film to be removed, and heating the film in the specific area. During heating film droplets are formed. The droplets move from the center of high intensity and evaporate. The resulting clean area is called *kerf*. The kerf is film-free.

Medicine   In medicine there is no technology that has advanced to the extent of the laser. Ruby lasers for years have been used for eye surgery. Argon ion lasers are useful as scalpels. The body skin fat and muscles are transparent to selective laser wavelengths, therefore cancer cells can be destroyed without harm to other parts of the body. Decayed areas of the teeth can be destroyed or removed without harm to healthy teeth. Laser beams have been made narrow enough to perform to microsurgery on red and white blood cells.

Alignment   As an alignment mechanism the laser has no equal. Placing a beam only microns in diameter several hundred feet to align an aircraft wing is significant. While all alignment is not this precise, the laser is indeed capable. Alignment of construction ditches is a case in point. Alignment in assembly of precision equipment is another. Lasers are used to align other lasers. An ultimate use in laser alignment is in military fire control and stability of precision direction-locating equipment.

IO

PHOTOMETRY/RADIOMETRY

PHOTOMETRY AND RADIOMETRY DEFINED

The Sciences Photometry is the science that deals with the measurement of visible light, that is, wavelengths from 390 to 770 nanometers depending on what reference you use. Radiometry is defined as the science that deals with the measurement of the entire optical spectrum including ultraviolet, visible, and infrared light. Wavelengths of light ranges from 0.005 microns to 4000 microns. (A micron is equal to 10^{-6} meter.) You may wish to review the section called "Optoelectronic Operational Spectrum" in chapter 1.

In the discussion of light measurement there are several important measurements of light necessary to understanding the science. These are flux, illumination, intensity, and luminance. Other terms are also important and will be defined in detail later in this chapter. In order to assess the major quantities involved in photometry and radiometry we must consider what it is we are going to monitor and how we can measure it accurately. This chapter will deal with the relationship of light-measurement quantities and the techniques involved in the measuring of these quantities.

PHOTOMETRIC AND RADIOMETRIC TERMS

General The relationship of photometric and radiometric terms is that they both deal with light energy or flux. Visible light energy is called *luminous flux* and is measured in *lumens*. Radiant light energy is called *radiant flux* and is measured in *watts*. Wavelengths of visible flux fall in the range of 390 to 770 nanometers in length. Wavelengths of radiant flux include all of the optical spectrum, that is, ultraviolet, visible, and infrared wavelengths. Luminous flux is related to radiant flux by a wavelength factor developed by the Commissioner Internationale de'Clairage (CIE) in France in 1924. (See figure 10–1.) The photopic eye re-sponse curve, as it is called, was developed from a sample of the population under daylight conditions. The response curve represents an average eye. The curve relates wavelength (λ) to relative luminosity or visual response and is called the luminosity function ($V\lambda$). The average eye sees best (is at its maximum efficiency) at 555 nanometers. The conversion factor for maximum efficiency of the average eye is 680 lumens per watt. Wavelengths other than 555 nanometers within the visible range must be converted to the luminosity function in relation to the CIE spectral response curve. Test equipment utilized in photometry must use sensors that match the CIE curve.

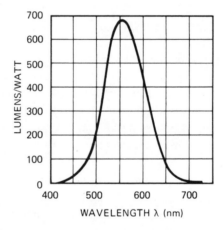

Figure 10–1 The CIE Standard Photopic Luminosity Function *(Courtesy United Detector Technology, Inc.)*

Light Sources Light energy is emitted in straight lines from one of two source types, the point source and the extended source (see figure 10–2). The point source is idealistically a single point; however, even an extremely small point of origin would be finite. The extended source of light energy illuminates a point in space from several directions.

The Solid Angle (ω) A sphere contains 4π steradians. The steradian is the unit measure of solid angles just as the radian is the unit measure of plane angles. A circle contains 2π radians. A sphere contains 4π steradians (sr). A

Figure 10–2 Light Sources (*Mike Anaya*)

hemisphere contains 2π steradians (sr). The solid angle ω is that angle with an apex in the center of the sphere and a radius r which subtends on the surface of the sphere. (See figure 10–3.)

Solid angle $(\omega) = \dfrac{S}{r^2}$

Solid angle $(\omega) = 1$ steradian (sr)   when $S = r^2$

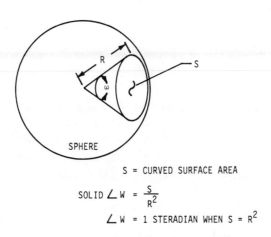

S = CURVED SURFACE AREA

SOLID $\angle$ W $= \dfrac{S}{R^2}$

$\angle$ W $= 1$ STERADIAN WHEN S = R^2

Figure 10–3 The Solid Angle (*Mike Anaya*)

Terms Relating to Photometry and Radiometry (See figure 10–4). The following terms are related to both optical measurement sciences.

Flux (ϕ)—Continuously changing light energy to, from, or through a surface
Incidence (E)—Flux per unit area incident (to) a surface
Exitance (M)—Flux per unit area exiting (from) a finite source
Intensity (I)—Flux per unit solid angle radiating from a finite source
Sterance (L)—Flux intensity per unit area leaving a light source

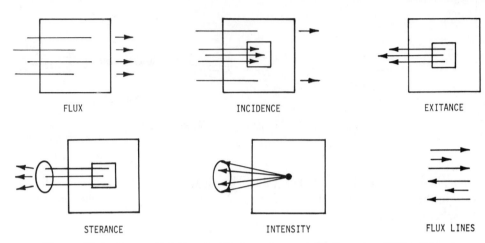

Figure 10–4 Terms Relating to Photometry and Radiometry (*Mike Anaya*)

TERMS RELATING TO PHOTOMETRY

The following terms are peculiar to measurement in the visible part of the optical spectrum, that is, the science of photometry. The reader will note that descriptive words in photometry such as prefixes will use the word luminous.

Candela (cd)   The basic unit of luminous intensity or flux per unit solid angle. The luminous intensity, in the direction of the normal, of a blackbody surface of 1/600,000 square meter in area, at the temperature of solidification of platinum under a pressure of 101,325 newton/meter^2. The term replaces the term candle, the obvious source. In 1948, the CIE redefined the term as previously stated.

Lumen (lm)   The basic unit of luminous flux. The lumen is equal to *one* candela (cd) of flux generated by a source point into a solid angle of *one* steradian (see figure 10–3).

The Lumen in Relation to the Solid Angle   You may recall that a sphere contains 4π steradians. As was just discussed, one lumen of flux is generated from a point source into a solid angle of one steradian. A point source of one candela (cd) of luminous flux will generate 4π lumens when the point source is generating flux in all directions (over the sphere). A point source of one candela (cd) of luminous flux will generate 2π lumens when the point source is generating flux in all directions (over the hemisphere). The luminous flux (lumens) is dependent on the luminous intensity (candelas) and the projection area.

Luminous Energy (Q_v)   Luminous energy is the generic term for luminous flux.

Luminous Flux (ϕ_v)   Luminous flux is the photometric measure for luminous energy. The SI unit of luminous flux is the lumen (lm).

Luminous Intensity (I_v)   The measure of luminous flux per unit solid angle from a point source. The unit of luminous intensity is lumens per steradian (lm/sr). The SI unit for luminous intensity is the candela (cd). The SI unit is recommended.

Luminance (B)   A measure of photometric brightness. The unit of luminance is lumens per steradian per square meter (lm/sr/m²). The SI measure for luminance is called luminous sterance (L_v).

Luminous Sterance (L_v)   A measure of photometric brightness. It is the flux per unit area emitting from a light source. The SI unit for luminous sterance is candelas per meter squared (cd/m²). Luminous sterance was previously called luminance.

Lambert (L)   A unit of luminous sterance equal to $1/\pi$ candela per centimeter squared ($L = 1\pi$ cd/cm²). Also, a foot-lambert is equal to $1/\pi$ candela per foot squared ($L = 1/\pi$ cd/ft²). Also a meter-lambert is equal to $1/\pi$ candela per meter squared ($L = 1/\pi$ cd/m²).

Foot-Lambert (fL)   A unit of luminous sterance equal to $1/\pi$ candela per foot squared (fL = $1/\pi$ cd/ft²).

Luminous Exitance (M_v)   The measure of flux exiting a finite surface. The SI unit for luminous exitance is lumens per square meter (lm/m²).

Illuminance (E)   The measure of flux per unit area incident to a surface. The unit of illuminance is lumens per square meter (1m/m²).

Luminous Incidence (E_v)   The measure of flux per unit area incident to a surface. The unit of luminous incidence is lux (lx).

Lux (lx)   The unit of luminous incidence in lumens per square meter (lm/m²).

Preferred SI Terms in Photometric Measurement

Luminous energy (Q_v)
Luminous flux (ϕ_v), in lumens (lm)
Luminous incidence (E_v), in lux (lx)

Luminous exitance (M_v), in lumens per square meter (lm/m^2)
Luminous intensity (I_v), in candelas (cd)
Luminous sterance (L_v), in candelas per square meter (cd/m^2)

The Inverse Square Law   This law of illumination states that the illumination of a surface due to a source point of light is proportional to the intensity of the source and inversely proportional to the square of the distance from the source to the surface. (See figure 10–5.) For example, the following table shows how much and how brightly a 1-lumen source will light an area from different distances.

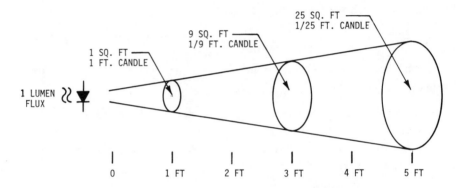

Figure 10–5   The Inverse Square Law (*Mike Anaya*)

Distance from Light Source	Illumination	Area
1 foot	1 footcandle	1 sq. ft.
3 feet	1/9 footcandle	9 sq. ft.
5 feet	1/25 footcandle	25 sq. ft.

The Cosine Law of Illumination   The cosine law of illumination states that the flux radiated from an extended source in a given direction varies with the projected area of the emitter or the receiver in a plane perpendicular to the direction of the flux. In the illustration, an angle is made between a normal line (perpendicular to the surface) and the incident flux lines. The angle is used to calculate the effective area presented to the incident flux. The effective area is area $\times$ cosine $\angle\theta$. The illumination in lumens per square foot is equal to the area $\times$ cosine $\angle\theta \times$ incident flux. See Figure 10–6.

NORMAL TO
SUFACE

INCIDENT
FLUX LINES

θ

EFFECTIVE AREA PRESENTED
TO INCIDENCE FLUX IS
AREA x COS $\angle\theta$

ILLUMINATION =
AREA x COS $\angle\theta$ x INCIDENT FLUX

Figure 10–6   The Cosine Law of Illumination (*Mike Anaya*)

TERMS RELATING TO RADIOMETRY

The following terms are peculiar to measurement in all parts of the optical spectrum. The reader will note that descriptive words in radiometry such as prefixes will use the word radiant.

Watt (W)   The basic unit of radiant flux. The watt is a measurement of power (energy) and relates to the photometric unit lumen (energy).

The Watt in Relation to the Solid Angle   You may recall that sphere contains 4π steradians. You may also recall that a solid angle is used as a parameter in the measurement of luminous intensity. Luminous intensity is measured in lumens per steradian (lm/sr). The parallel radiant intensity is measured in watts per steradian (W/sr).

Radiant Energy (Q_e)   Radiant energy is the generic term for radiant flux.

Radiant Flux (ϕ_e)   Radiant flux is the radiometric measure for radiant energy. The SI unit of radiant flux is the watt (W). Radiant flux is also called radiant power.

Radiant Intensity (I_e)   The measure of radiant intensity per unit solid angle from a point source. The unit of radiant intensity is watts per steradian (W/sr).

Radiance (N)   Flux per unit solid angle per area of emitting surface. The unit of radiance is watts per steradian per meter squared (W/sr/m²).

Radiant Sterance (L_e)   Flux per unit solid angle per area of emitted surface. The unit of radiant sterance is watts per steradian per square meter (W/sr/m²). Radiant sterance was previously called radiance.

Emittance (W)   Flux per unit area exiting from a finite source. The unit of emittance is watts per square meter (W/m²).

Radiant Exitance (M_e)   The measurement of flux per unit area exiting from a finite source. The unit of radiant exitance is watts per square meter (W/m²). Radiant exitance was previously called emittance.

Irradiance (H)   Flux per unit area incident to a surface. The unit of irradiance is watts per steradian per square meter (W/sr/m²).

Radiant Incidence (E_e)   The measurement of flux per unit incident to a surface. The unit of radiant incidence is watts per square meter (W/m²). Radiant incidence was previously called irradiance.

Preferred SI Terms in Radiometric Measurement

Radiant Energy (Q_e)
Radiant Flux (ϕ_e), in watts (W)
Radiant Incidence (E_e), in watts per square meter (W/m²)
Radiant Exitance (M_e), in watts per square meter (W/m²)
Radiant Intensity (I_e), in watts per steradian (W/sr)
Radiant Sterance (L_e), in watts per steradian per meter squared (W/sr/m²)

THE PHOTOMETER/RADIOMETER

A photometer/radiometer is a device that measures the intensity and sterance of a light beam. Any of the basic photometer/radiometer units consist of a light detector, an amplifier, a function selector, and either an analog or a digital readout/display. Light from some source, such as an LED or a laser is exposed to the photosensitive surface of a light detector. The light detector in modern units is usually a PIN silicon diode. The purpose of the PIN diode is to convert the light energy to electrical current. The current is proportional to the intensity of the light beam and varies with its wavelength. This wavelength must be carefully considered when recording optical-power measurements. The variance is named *spectral response*. The light energy in the form of electrical current is directed to amplifier. The amplifier's gain is controlled by a range switch. A function selector selects the mode in which the meter will operate. The amplifier output is then directed into an ammeter meter movement or a digital display. The position of the meter movement or the displayed numerals is directly proportional to the intensity of the light at the power meter detector.

Light Suppression   When monitoring a light source, it is often necessary to utilize an external light attenuator to suppress background light. Background light, also called *ambient light,* may cause errors in power measurement especially with low power measurements. The ambient light may be handled in several ways. Before measuring a desired light source, a measurement of the ambient light is made, then the total source and ambient light is measured. Simple subtraction of ambient-light measurement from the total provides the source-light measurement. A second method is built into some power instruments. A calibration adjustment zeros the power meter, which in effect calibrates out the ambient light. A third method is to utilize an ambient light shade. The shade is placed over the detector and allows only the source light to reach the detector.

Light Attenuation   Just as ambient light affects the reading of an optical power meter at low power, the opposite situation of high power is a problem. When measuring very high power, the optical range of a power meter may be exceeded without the operator being aware of it. In this event extreme damage to the detector and possibly the meter movement could result. To alleviate this problem an attenuator is placed in front of the detector in the path of the light beam under measurement. This limits the beam power reaching the detector. The attenuator is made of diffusing material and scatters the beam within its walls. Part of the beam is absorbed by the attenuator's surfaces. A small amount falls on the detector surface and is measured. The amount of light energy falling on the detector can be changed by adjusting the distance between the detector and the attenuator diffusing material. Shortening the distance provides less attenuation, lengthening the distance provides more attenuation. When external attenuators are used, the meter reading must be adjusted in relation to the attenuation factor.

Photometric Filtering   Optical power meters are often used to measure light energy that is similar to that seen by the human eye. To measure this light, the meter must respond to it the same way the human eye would. To accomplish this a photometric filter is placed in front of the light-source detector. This causes the meter to react the same as the eye would react to the light source. Figure 10–7 illustrates the CIE photopic curve in relation to photodiode filters. The photopic (human eye) curve is the solid line. Two PIN photodiode/filters are compared. These curves represent the wavelength response of the human eye and the filters.

Figure 10–7 The Photometric Filter Meter Correction Curve (*Courtesy United Detector Technology, Inc.*)

The reader will note that the difference is very slight. Most excellent power meters will have a selector switch that provides for photometric measurements.

Radiometric Filters Optical power meters are often used to measure light energy that is beyond the range of visible wavelengths (not photometric). To measure these optical wavelengths a radiometric filter is used (see figure 10–8). The filter is placed over the detector and absorbs selected wavelengths of light while allowing other wavelengths to be seen by the detector with little or no absorption. When the filter is in place the meter responds to all wavelengths from approximately 400 to 1000 nanometers in the same level.

Illuminance/Irradiance and Luminance/Radiance Measurements Illuminance is the measurement of the luminous flux density (lumens/unit area) at a receiving surface as seen by a photopic sensor. The typical measurement unit in

Figure 10–8 The Radiometric Filter Meter Correction Curve (*Courtesy United Detector Technology, Inc.*)

English-speaking countries is footcandle (lumen/sq. foot) and in most other countries is lux (lumen/sq. meter).

Irradiance is the radiant flux density (watts/unit area) as measured with a flat response sensor over given wavelengths. Illuminance/irradiance measurements are accomplished by setting the function selector at footcandles (illuminance) or microwatts (irradiance), screwing the appropriately marked filter (photometric or radiometric) onto the head assembly and directing the head at the desired light source.

Luminance (commonly termed brightness) is what the eye perceives, as specified by the C.I.E. curve. Luminance is luminous intensity of the source per unit source area projected in the direction of view (candelas per unit area). Radiance is the radiometric equivalent of luminance (watts/unit area of source/ unit solid angle).

Light Flash Measurements   Measurements can be made radiometrically in microjoules (microwatts-seconds) or photometrically in talbots (footcandle-seconds). Most photometers/radiometers are capable of measuring light flashes. Those that are capable accomplish this task in the following manner. Radiometric energy measurements are accomplished by attaching the radiometric filter to the sensor head, setting the function selector at microwatts and the range selector to the desired energy scale. Meter readings are then definable in microwatts-seconds. Photometric flash measurements are accomplished by attaching the photometric filter to the sensor head, setting the function selector at footcandle and the range selector on the desired energy scale.

Densitometry Measurement   Most photometers/radiometers can be used as a precision transmission densitometer for measuring the optical density of film, neutral density filters, colored-glass filters, and coated optics. Direct percent transmission readings can be obtained when the sample is inserted between the light source and detector head. Densitometric measurements can be made with a photopic detector response, spectrally flat radiometric response, or isolated wavelength response by the use of detachable filters.

A Typical Portable Optical Power Meter   The UDT–40X Optometer (see figure 10–9) measures radiant power, radiant energy, illuminance, luminance, and integrated luminance over seven decades (logarithmically over six decades). A 1 cm² PIN silicon diode is used as the basic sensor. Two computer-designed, subtractive filters are provided for radiometric and photometric measurements. The radiometric filter provides a flat spectral response from 450 to 910 nanometers. The photometric filter yields a spectral response comparable to the human eye. The relative reading function position "D" allows the user to set full scale at any value desired. The energy ranges integrate input light power from durations of 10 nanoseconds to several minutes. A manual reset button discharges the stored charge and resets the meter to zero. A logarithmic readout scale covers six decades of input light power without changing range.

Low-noise operational amplifiers are used to increase photodiode signal current to display levels. A taut band, mirrored scale, 4-inch meter movement displays the output. All the electronics are contained in a metallic shielded enclosure. The combination photodiode/operational amplifier is extremely stable and maintains its calibration independent of temperature and light-level extremes over many months. The instrument is supplied calibrated radiometrically and photometrically.

Figure 10–9 The UDT 40X, Portable Radiometer/Photometer (*Courtesy United Detector Technology, Inc.*)

A Typical Laboratory Photometer/Radiometer The UDT–11A Photometer/Radiometer measures radiant power, radiant energy, luminance, illuminance, and spectral radiance (see figure 10–10). This model is designed as a laboratory standard with accuracies of 1 percent throughout all ranges. The meter movement has an accuracy of 0.05 percent, ± 1 digit-reading solution. Nine decades of light level can be measured from the lowest to the highest sensitivity. Construction of the model is of solid-state devices. The instrument can measure currents of 10^{-12} amperes. The range of the instrument is from 10^{-3} to 10^{3} microwatts. Its sensitivity is 10^{-11} watts, 10^{-5} footcandles, and 10^{-4} foot-lamberts. The unit can be used as a standard photometer/radiometer or with different heads has a variety of uses. Some of these are as a telephotometer, microphotometer, spectro-radiometer, and LED instrument.

In figure 10–10, the UDT–11A is used along with a microphotometer. The microphotometer is basically a microscope with a 0.100-inch detector located as a target in the focal plane of the objective. The field of view is precisely defined by the detector area subtended at the objective. The objective resolves a spot size of 0.030 inch at a measuring distance of 1 inch. A computer-selected filter located in the detector filter matches the C.I.E. (human eye) response to within ±2 percent of the total integrated area under the spectral curve. The microphotometer can be used as a microradiometer by simply removing the photometric filter.

Figure 10–10 The UDT 11A Photometer/Radiometer (*Courtesy United Detector Technology, Inc.*)

THE LASER POWER METER

The purpose of a laser power meter is to measure the output power of a laser directly. It accomplishes this by coupling an extremely high-quality PIN diode with an integrating sphere. Figure 10–11 illustrates the integrating sphere and how it is set up to perform laser power measurements. In A of the figure a cross section of an integrating sphere is shown. The integrating sphere is accurately machined with a diffusely reflecting internal surface. An incoming light beam is dispersed through multiple diffuse reflections into a uniform intensity over the entire inner surface of the sphere. A partially transmitting diffuser, positioned on this surface and directly in front of an optical detector, accurately detects an exact percentage of the total input light. This energy level is then calibrated to accurately read the total input light beam.

Figure 10–11 Integrating Sphere and Test Setup for Measuring Laser Power (*Courtesy United Detector Technology, Inc.*)

In figure 10–11, the test set up with the laser is shown. The integrating sphere is placed on a rack in line with the laser beam under test. An accurate power readout requires only that the entire laser beam being measured be directed into the input port. The integrating sphere does not have the interference problems encountered with the use of density filters. The detector is mounted on the top of the integrating sphere. Leads are connected from the detector to a detector-head input jack on the forward side of the power meter. The power meter indicates the laser power on a digital nixie display.

Measurement of laser power requires exacting safety precautions. These standards vary for individual wavelengths but for the visible region of 440 to 700 nanometers, Class I lasers must not permit human access to radiation in excess of 0.4×10 watts, Class II lasers must be below 1×10 watts, Class III lasers below 0.5 watts, and Class IV lasers are above .5 watts. In all cases various safety standards and precautions must be adhered to and displayed. An optional 7-millimeter diameter aperture provides a direct power reading in accordance with proposed government input-aperture specifications.

A Typical Laser Power Meter   A typical laser power meter is the United Detector Technology (UDT) Model 21 shown in figure 10–12. The power meter is used along with the UDT2500 integrating sphere and the UDT223 detector filter assembly to measure laser power. The meter has a flat calibrated response over a wide range of laser wavelengths (450 to 900 nanometers). The meter can measure laser power over 12 full decades from 10^{-10} watts to 100 watts of input power. A digital readout requires no calibration curves. An offset adjustment provides a convenient means of meter zeroing and suppression of minor back-

Figure 10–12 The UDT Model 21 Laser Power Meter (*Courtesy United Detector Technology, Inc.*)

ground radiation. Low-level power can be measured by removing the detector from the integrating sphere and placing it in line with the laser beam. Lower-level power is 10^{-10} watts to 1 milliwatt.

LED Display Measurement System The United Detector Technology Model 111L is representative of an LED measurement system. It is pictured in figure 10–13. The purpose of the LED display measurement system is to measure the total flux of green, yellow, or red LED displays in terms of luminous intensity. Single- or multiple-segment LEDs are mounted on a plate. The LEDs are energized by an external power supply. It is not necessary for the LED segments to have any exact alignment. The LED plate is mounted on a drawer that slides into the measurement module.

Figure 10–13 LED Display Measurement System (*Courtesy United Detector Technology, Inc.*)

The measurement module consists of the LED enclosure and the detector head. The detector head is a black anodized optical cylinder containing detector, photometric filter, lens, and baffles. It is designed to position the detector at a fixed distance from the LED being measured. The sensor is a critically processed instrument-quality silicon detector with a circular active area of 1.0 cm^2. This high-performance detector is specially selected for long-term stability, high responsivity, low-noise uniformity, and linearity. Just below the detector is a red filter. The photometric filter is a computer-designed multilayered optical-quality glass filter. This filter corrects the detector's spectral response to that of the CIE curve and provides exceptional match for all visible LED's. The lens defines the field of view of the system and focuses the energy emitted from the LED onto the detector. A series of baffles within the blackened cylinder eliminates the possibility of stray light causing erroneous measurements. The LED enclosure has a positioning module with provision for detachable mounting plates.

The control module contains the readout, electronics, power supply, and all necessary controls. A $3\frac{1}{2}$-digit numerical display gives immediate direct readout in candelas. Direct-readout calibration channels are controlled by a four-position switch located on the rear panel. Each of the LED types, green (565 nanometers), yellow (585 nanometers) and red (665 nanometers), has its own dedicated calibration channel. A fourth channel is available for any other calibration required at the user's option. The particular channel in use is indicated on the control panel by a lighted "annunciator" of the same color as the LED being measured. Readings are displayed in absolute units and require no further reduction. A range switch provides system sensitivity over seven full-scale decades. A high-resolution full-scale offset control permits all stray light to be offset to zero. Outputs on the rear panel include analog and TTL compatible BCD.

RADIOMETER/PHOTOMETER ACCESSORY EQUIPMENT

Microscope Accessory   The microscope accessory can measure brightness or radiance of small objects ranging from say 0.15 to 0.76 millimeters with a circular field of view.

Spot Brightness Accessory   The spot brightness accessory is a reflex viewing accessory used with an objective lens to provide a variety of working distances and resulting image sizes.

Detector Holder and Aperture Set   The detector holder and aperture set provides a convenient means of varying detector area without changing detector response.

Steradian Shade To fix the field of view to a specific geometry, the steradian shade should be used. The geometry is fixed at say 0.155 steradians. This permits the measurement of surface brightness (luminance or radiance) in terms of foot-lamberts, candelas per square meter, or microwatts per square centimeter steradian.

Light Shade The light shade is used in conjunction with detector/filter assemblies to limit the detector field of view in degrees thereby minimizing off-axis ambient light effects.

LED Shade The LED shade is a baffled tube that accurately defines the field of view, necessary when measuring LEDs.

Neutral Density Filters The purpose of the neutral density filter is to extend the maximum power measurement capability of radiometers and photometers.

Laser Beam Spreader The beam spreader extends laser power measurements capability.

Laser Integrating Sphere The laser integrating sphere provides constant attenuation for high-power laser measurements.

LED Integrating Sphere The LED integrating sphere facilitates laser power measurements. The sphere extends power measurements capability.

Laser Attenuator The laser attenuator provides constant attenuation for higher power laser measurements.

Photometric Filter When it is desired to measure light as the human eye would perceive it, the photometric filter should be used.

Cosine Diffuser The cosine diffuser is employed when making measurements of radiation at a surface.

Radiometric Filter The radiometric filter is used to provide a flat response when measuring radiation other than photometric.

Cosine Receptor The cosine receptor is employed to measure incident light over a wide angle.

Narrow Bandpass Filters   Optical glass interference filters with half-power bandpass at popular laser wavelengths.

Fiber Optic Probes   Probes used to make measurements in inaccessible areas. The probe is a flexible light guide constructed of glass-clad optic fibers sheathed in PCV with stainless steel end fittings.

Objective Lens   Lens used to provide specific spot sizes.

Close-Up Objective   Provide a variable zoom at a specific degree viewing angle.

Lumilen   Lens used with a detector head for measurement of surface brightness.

Temperature Filters   Filters used with a radiometer/photometer to determine color temperatures in Kelvins of incandescent sources.

GLOSSARY

Acceptance angle The critical angle, measured from the core centerline, above which light will not enter the fiber. It is equal to the half-angle of the acceptance cone.

Acceptance pattern The acceptance pattern of a fiber or fiber bundle is a curve of input radiation intensity plotted against the input (or launch) angle.

Active medium A collection of atoms excited to produce a population inversion.

Angstrom ($\overline{A}$) A wavelength value equal to $1/10,000$ micron or 10^{-10} meter.

Antinodes The maximum amplitude point in standing waves.

Atomic lifetime The amount of time that an atom stays in an excited state before spontaneously falling to another, lower energy state.

Beam A slender shaft or stream of light or other radiation.

Beamwidth (θ_0) The difference between the angles at which radiant intensity $I(\theta)$ is 50 percent (unless otherwise stated) of the peak value I_p.

Branch That portion of a cable or harness that breaks out from and forms an arm with the main cable or harness run.

Breakout The point where a branch meets and merges with the main cable or harness run or where it meets and merges with another branch.

Brewster window Glass windows placed at exact angles on the ends of a plasma tube.

Brightness Radiance. The amount of power radiated by a laser beam.

Buffer A protective material that covers and protects a fiber. The buffer has no optical function.

Bundle A number of fibers grouped together.

Bundle jacket The material that is the outer protective covering common to all internal cable elements.

Cable A jacketed bundle or jacketed fiber in a form that can be terminated.

Cable assembly A cable that is terminated and ready for installation.

Cable core The portion of a cable contained within a common covering.

Cable jacket The material that is the external protective covering common to all internal cable elements.

Cable or harness run That portion of a branched cable or harness where the cross-sectional area of the cable or harness is the largest.

Circularly polarized   A wave is said to be circularly polarized when the electrical field traces out a circle in a plane perpendicular to the direction of wave propagation.

Cladding   A sheathing or cover of a lower refractive index material intimately in contact with the core of a higher refractive index material. It serves to provide optical insulation and protection to the total reflection interface.

Cladding mode stripper   A material applied to the fiber cladding that provides a means for allowing light energy being transmitted in the cladding to leave the cladding of the fiber.

Coherence   Light waves which are in-step, in-phase, and monochromatic.

Continuous wave (CW)   A laser whose output is continuous with no interruptions.

Core   The high refractive index central material of an optical fiber through which light is propagated.

Corpuscle theory   Theory of light in seventeenth century that suggested light consists of corpuscles that had an internal vibration of their own or in some way were controlled by waves or vibrations of the medium (such as air) through which they travel.

Coupler   An optical device used to interconnect three or more optical conductors.

Coupling loss   The total optical power loss within a junction, expressed in decibels, attributed to the termination of the optical conductor.

Critical angle   The maximum angle at which light can be propagated within a fiber. Sine critical angle equals the ratio of the numerical aperture to the index of refraction of the fiber core.

Crosstalk   Measurable leakage of optical energy from one optical conductor to another.

Crystalline   Solid material that has a lattice crystal structure such as the ruby.

CVD fiber (chemical vapor deposition)   A process by which a heated gas produces an oxide deposit to fabricate a glass fiber preform. The deposited glass becomes the core.

Decibel   The standard unit for expressing transmission gain or loss and relative power levels. A decibel equals ten times the log of the ratio of the power out to the power in. ($dB = 10 \log_{10} P_0/P_1$)

Detector   A device that converts optical energy to electrical energy, such as a PIN photodiode.

Diffraction   Breaking up of light rays into bands when a ray is deflected by an opaque object or a slit.

Diffusion   A scattering of light rays. A softening of light.

Directionality   Not spread out in all directions. A narrow cone projected in one direction.

Dispersion   The "spreadout" or "broadening" of a light pulse as it propagates through the optical conductor. Dispersion increases with length of conductor and is caused by the difference in ray path lengths within the fiber core.

Divergence   Spreading out of a light beam.

Electroluminescence   Electrical energy application to light-sensitive material causing emission of light (radiant) energy.

Emission   Ejection of electrons from a surface by radiation.

Emission theory   Theory of light developed about 500 B.C. by Greek thinkers. The theory suggested that all objects emit something which, when entering the eye, was caused by some part of the eye, therefore the brain could see.

End finish   Surface condition at the optical conductor face.

Exit angle   The angle between the output radiation vector and the axis of the fiber or fiber bundle.

Excitation mechanism   Source of energy which excites the atoms in an active medium.

Extended source   Light projecting from

several points in space to illuminate a point from many directions.

Extinction A measure of how well a beam is polarized. It is the ratio of the power transmitted through an aligned polarizer and power transmitted through a polarizer rotated 90°.

Feedback Coupling of energy from output back into an active medium.

Fiber A single discrete optical transmission element usually comprised of a fiber core and a fiber cladding.

Fiber buffer The material that surrounds and is immediately adjacent to a fiber that provides mechanical isolation and protection. Note: Buffers are generally softer materials than jackets.

Fiber bundle A consolidated group of single fibers used to transmit a single optical signal.

Fiber cable A cable composed of a fiber bundle or single fiber, strength members, and a cable jacket.

Fiber cladding That part of a fiber surrounding the core of the fiber and having a lower refractive index than the core.

Fiber core That part of a fiber having a higher refractive index than the cladding that surrounds it.

Fiber jacket The material that is the outer protective covering applied over the buffered or unbuffered fiber.

Fiber optics (FO) A general term used to describe the function where electrical energy is converted to optical energy, then transmitted to another location through optical transmission fibers, and converted back to electrical energy. While the bulk of applications is in digital transmission data, fiber optics also can be applied to analog systems.

Fiber sheath A general term used variously to mean cladding, buffer, or jacket. Not to be used in military FO specifications.

Finite Measurable.

FO An abbreviation for fiber optics.

Footcandle Measure of intensity of illumination at all points of an illuminated surface one foot from a one candle source.

Frequency The number of times a wave occurs in a period of time.

Fresnel reflection losses The losses that are incurred at the optical conductor interface due to refractive-index differences.

Fringe A pattern of dark and light bands created by shining a monochromatic beam of light on a surface such as in thin films.

Fundamental mode A transverse electromagnetic mode (TEM) that oscillates in one mode only.

Gas loss A power loss, expressed in decibels, due to the deviation from optimum spacing between the ends of separable optical conductors.

Graded-index fiber A fiber whose index of refraction decreases with increasing radial distance from the center of the core.

Ground state The lowest energy state of an atom.

Harness A construction in which a number of multiple-fiber cables or jacketed bundles are placed together in an array that contains branches. A harness is usually installed within equipment or airframe and mechanically secured to that equipment or airframe.

Harness assembly A harness that is terminated and ready for installation.

Hertz (Hz) Cycles per second of a radiated frequency.

Heterochromatic Of, having, or consisting of many colors.

Index-matching materials Materials used in intimate contact between the ends of optical conductors to reduce coupling losses by reducing Fresnel loss.

Incoherent Light waves that are not in-step, not in-phase, and nondirectional.

Index of refraction The ratio of the speed of light in a vacuum to the speed of light in a material. When light strikes the surface of a transparent material, some light is reflected while some is refracted.

Infinite Lacking limits, extending beyond the bounds of measurement or comprehension.

Infrared Band of light wavelengths too long for response by a human eye.

Kilohertz One thousand cycles per second of a radiated frequency.

Lambda (λ) Greek letter depicting wavelength. Mathematcally $\lambda = c/f$, where c is velocity in meters (300,000,000 meters per second) or (186,000 miles per second) and f equals frequency.

Lambertian A radiance distribution that is uniform in all directions of observation.

Laser Light amplification by stimulated emission of radiation.

Lateral loss A power loss, expressed in decibels, due to the deviation from optimum coaxial alignment of the ends of separable optical conductors.

Launch angle The angle between input radiation vector and the axis of the optical conductor.

Linearly polarized A wave is said to be linearly polarized when the electrical field oscillates along a line in a plane that is perpendicular to the direction of wave propagation.

Lumen Measure luminous flux of one square foot of spherical surface one foot from a one candle source.

Maser Microwave amplification by stimulated emission of radiation.

Megahertz One million cycles per second of a radiated frequency.

Meridial rays The rays of light that propagate by passing through the axis of the fiber and travel in one plane.

Metastable state Energy states of atoms that have extremely long atomic lifetimes.

Micron (μ) A wavelength value equal to 10,000 angstroms or 10^{-6} meter.

Microwave Electromagnetic waves extending from 300 MHz to 300,000 MHz.

Monochromatic Of, having, or consisting of one color.

Multiple-bundle cable assembly A multiple-bundle cable terminated and ready for installation.

Multiple-fiber cable assembly A multiple-fiber cable that is terminated and ready for installation.

Nodes The zero amplitude points in standing waves.

Numerical aperture (NA) The characteristic of an optical conductor in terms of its acceptance of impinging light. The sine of the acceptance cone half-angle equals the square root of the difference of the square of the index of refraction of the fiber core minus the square of the refractive index of the fiber cladding.

Numerical aperture, NA (10% intensity) The numerical aperture of a fiber or bundle is defined by: NA (10% intensity) = sine θ' where θ' is the angle where the measured intensity of radiation is 10 percent of the maximum measured intensity when plotting either the acceptance or radiation pattern.

Numerical aperture, NA (materials) The numerical aperture of a fiber or bundle is defined by: NA = $(n1^2 - n2^2)^{1/2}$ where n1 and n2 are the fiber core and cladding refractive indices, respectively, for step-index fibers. For graded-index fibers n1 is the maximum index in the core and n2 is the minimum index in the clad.

Numerical aperture, NA (90% power) The numerical aperture of a fiber or bundle is defined by: NA (90% power) = sine θ where θ is the angle between the axis of the output cone of light and the vector coincident with the surface of a cone that contains 90 percent of the total output radiation power, or where θ is the angle between the axis of the input cone of light and the vector coincident with the surface of a cone which contains 90% of the total input radiation power.

Opaque Material that blocks light rays.

Optical conductors Materials that offer a

low optical attenuation to transmission of light energy.

Optical spectrum Wavelengths that operate in the light and sight spectrum.

Optics Branch of physics dealing with light and vision.

Optoelectronics Branch of electronics dealing with light.

Packing fraction (PF) The ratio of the active core area of a fiber bundle to the total area at its light-emitting or receiving end.

Panchromatic Sensitive to light of all colors.

Peak radiant intensity (I_p) The maximum value of radiant intensity, I.

Peak wavelength (λ_p) The wavelength at which the radiant intensity is a maximum. Unit: nanometers (nm).

Period The length of time designated to count the number of wave cycles. Usually the time is in seconds.

Phase Value of a wave in relation to another or to a specific time.

Photoconduction Change of resistance in a material when it is exposed to radiation.

Photoelectric Electrical effects caused by light or other radiation.

Photoelectrons Electrons that absorb light energy or photons.

Photoemission Light releasing electrons from a surface.

Photon Tiny pulses of light energy. An atom stimulated in an excited state emits photons of energy. A quantum of light energy.

Photovoltaic Generation of voltage with the use of dissimilar materials and light-sensitive material in response to radiation.

Plasma Ionized gas.

Plastic-clad silica fiber (PCS) A fiber composed of a silica-glass core with a transparent plastic cladding.

Point source Light projecting from one point in all directions.

Population inversion Movement of an

atom from a ground energy state to an excited energy state.

Power density Irradiance.

Pulsed A laser whose output power has predetermined changes with time.

Quantum Unit of light energy dependent on wavelength.

Radiance The radiant flux per unit solid angle and per unit surface area normal to the direction considered. The surface may be that of a source, detector, or any other surface intersecting the flux.

Radiant energy Energy generated by the frequency of radiation of electrons. The energy of light.

Radiant intensity ($I = d\phi/d\omega$) The radiant power per unit solid angle ω in the direction considered measured in watts per steradian (W/sr).

Radiant power (ϕ) The time rate of flow of electromagnetic energy, measured in watts (W).

Radiant power ratio R is defined as the radiant power ratio $\phi 2/\phi 1$, where $\phi 1$, and $\phi 2$ are the measured power before and after specimen conditioning, respectively.

Radiation Process by which energy in the form of rays is sent through space from atoms and molecules as they undergo internal change.

Radiation pattern The radiation pattern is a curve of the output radiation intensity plotted against the output angle.

Ray A line of light extending between two points.

Ray angle The angle between a light ray and a reference line or plane.

Receiver optical An electro-optical module that converts an optical input signal to an electrical output signal.

Refraction The bending of a light ray when the ray moves through the intersection of two mediums such as air to water.

Repeater A device that converts a received optical signal to its electrical equivalent, reconstructs the source signal format, amplifies, and reconverts to an optical output signal.

Spatial coherence Phase correlation of two different points across a wave front at a specific moment in time.

Selenium Metal whose resistance is effected by light radiation.

Skew rays Rays of light that do not propagate through the axis of the fiber.

Source The source of radiant energy, such as a light-emitting diode (LED).

Spectral bandwidth (λ_{BW}) The difference between the wavelengths at which the radiant intensity $I(\lambda)$ is 50 percent (unless otherwise stated) of the peak value I_p.

Spectrum Ranges of frequencies or wavelengths.

Splice Nonseparable junction joining optical conductor to optical conductor.

Spontaneous emission When an atom falls from an excited state to a ground state and in doing so emits a photon of energy, the photon is said to have been spontaneously emitted.

Sputtering Atoms jarred loose from a cathode by bombardment of ions in a plasma tube.

Standing waves A pattern created by waves reflecting back and forth within a laser cavity. These waves have the appearance of standing still if they are of the same frequency, amplitude, and moving in opposite directions.

Step index fibers A fiber in which there is an abrupt change in refractive index between the core and cladding along a fiber diameter.

Stimulated absorption When an atom absorbs a photon and moves to a higher energy state.

Stimulated emission When an external source of power causes an atom to move from an excited state to a lower state of energy.

Tactile theory Theory of light accepted by Greek thinkers about 500 B.C. It asserted that eyes sent out invisible antennae and were thus able to feel or sense those things too distant to touch.

Temporal coherence Phase correlation of waves at a point in space at two instants of time.

Translucent Material that allows light to partially pass through.

Transmission mirror Mirror on the output of a laser that partially transmits the light beam and reflects the rest of the light back into the active medium.

Transmitter optical An electro-optical module that converts an electrical input signal to an optical output signal.

Transparent Material that clearly allows light to pass through.

Transverse wave A wave whose amplitude is perpendicular to its direction of propagation.

Uniform lambertian A lambertian distribution that is uniform across a surface.

Ultraviolet Band of light wavelengths too short for response by a human eye.

Unpolarized A wave is said to be unpolarized when the electrical field oscillates randomly in a plane perpendicular to the direction of propagation.

Velocity Rate of change of position in relation to time.

Visible light Electromagnetic wavelengths that can be seen by the human eye ranging from 380 nanometers to 770 nanometers.

Wave The motion of light as it travels through space.

Wave crests The part of a light wave that has maximum vibration.

Wave front The forward side of a light wave.

Wavelength The amount of space occupied by the progression of an electromagnetic wave.

Wave trough The part of a light wave that has minimum vibration.

BIBLIOGRAPHY

BOOKS

Allen, W. B., *Fiber Optics.* Plenum Press, London, 1979.

Clarricoats, P. J. B., Ph.D., *Optical Fiber Waveguides.* Peter Peregrinus LTD, Institution of Electrical Engineers, Southgate House, Stevenage, Herts SG1 1Hq. England, 1975.

Gloge, Detlaf, ed., *Optical Fiber Technology,* The Institute of Electrical and Electronics, Inc. New York, 1976.

Midwinter, John E., *Optical Fibers for Transmission,* John Wiley & Sons, New York, 1979.

PAMPHLETS

Anderson, John D., "Fiber Optic Data Bus," Northrup Corp., Hawthorne, California.

Bark, P. R., and Wey, R. A., "Field Splicing of Optical Cables on Fiber Assessment with Optical Time Domain Reflectometry," Siecor Optical Cables Inc., Horseheads, New York.

Barton, David M., "Driving High Level Loads with Iso-Lit Opto-Isolators," Litronix Inc., Cupertino, California.

Barton, David M., "More Speed from Iso-Lit Optical Isolators," Litronix Inc., Cupertino, California.

Bielowski, W. B., "Low Loss Waveguide Communication Technology," Corning Glass Works, Corning, New York.

Bishop, O. L., and Smith, J. C., "Installation of a Fiber Optic System in an Electrical Power Station," ITT Telecommunications, Raleigh, North Carolina.

Bloom, Arnold L., "C. W. Pumped Dye Lasers," Coherent Inc., Palo Alto, California.

Borsuk, Les, "Introduction to Fiber Optics," ITT Cannon Electric, Santa Ana, California.

Borsuk, Les, "Fiber Optic Interconnections," ITT Cannon Electric, Santa Ana, California.

Bowen, Terry, "Low Loss Connectors for Single Optical Fibers," AMP Inc., Harrisburg, Pennsylvania.

Ebhardt, C. A., "A T1 Rate Fiber Optic Transmission System for Electrical Power Station Communications," ITT Transmission Division, Raleigh, North Carolina.

Ebhardt, C. A., Edward, A. K., and Magner, R. G., "A T3 Rate, Repeatered Fiber Optic Transmission Systems," Raleigh, North Carolina.

Eppes, T. A., Goell, J. E. and Gallenberger, R. J., "A Two Kilometer Optical Fiber Digital Transmission System for Field Use at 20 Mb/s," ITT Electro-Optical Products Division, Roanoke, Virginia.

Fenton, Kenneth, "Fiber Optic Connectors," ITT Cannon Electric, Santa Ana, California.

Martinet, R. L., "A T3 Rate, Repeatered, Fiber Optic Transmission System," ITT Electro-Optical Products Division, Roanoke, Virginia.

Miskovic, Edward J., "Fiber Optic Data Bus," ITT Cannon Electric, Santa Ana, California.

Nemhauser, Robert I., Alexander, Gil, and Duda, Richard, "Radiometry and Photometry Once Over Lightly," United Detector Technology, Inc., Santa Ana, California.

Olszewski, J. A., Foot, G. H., and Huang, Y. Y., "Development and Installation of an Optical-Fiber Cable for Communications," General Cable, Union, New Jersey.

Schumacher, William L., "Fiber Optic Connector Design to Eliminate Tolerance Effects," AMP Inc., Cherry Hill, New Jersey.

Smith, George, "Applications of Opto-Isolators," Litronix Inc., Cupertino, California.

Smith, George, "LEDs and Photometry," Litronix Inc., Cupertino, California.

Smith, George, "Multiplexing LED Displays," Litronix Inc., Cupertino, California.

Takagishi, Dave, "Applying the DL-1416 Intelligent Display," Litronix Inc., Cupertino, California.

Wey, Robert A., and Cole, James A., "Optical Fibers, Cables, and Links for Computer Systems Applications," Siecor Optical Cables Inc., Horseheads, New York.

Wheeler, John Post, "Design and Manufacture Considerations in Commercial Ion Lasers," Palo Alto, California.

INDEX